AF703004

Quantum Horizon

Mastering the Next Era of Advanced Computing

Laura Lee

© 2024 Laura Lee. All rights reserved.
Published by Lauxon Publishing, LLC
No part of this book may be reproduced, distributed, or transmitted in any form or by any means, including photocopying, recording, or other electronic or mechanical methods, without the prior written permission of the publisher, except in the case of brief quotations embodied in critical reviews and certain other noncommercial uses permitted by copyright law.

For permission requests, please contact:
Email: info@Lauxonpublishing.com
Website: www.Lauxonpublishing.com

Section 1: Foundations of Quantum Mechanics

Section 2: Building Quantum Computers

Section 3: Quantum Error Correction

Section 4: Quantum Algorithms

Section 5: Quantum Machine Learning

Section 6: Quantum Cryptography and Security

Section 7: Quantum Software and Programming

Section 8: Applications in Science and Industry

Section 9: Challenges and Future Directions

Section 10: Visionary Concepts and Theoretical Horizons

Prologue:

Into the Quantum Era

There are moments in human history when the rules change.

The Industrial Revolution redefined work. The Digital Age transformed communication. And now, quietly but unmistakably, a new revolution is taking shape—one that doesn't just change how we compute, but what we can compute.

This is the dawn of the **quantum era**.

For over a century, quantum mechanics has been a strange, abstract pillar of theoretical physics. It defied common sense, challenged the classical worldview, and unsettled even the greatest minds of the 20th century. But today, its weirdness is no longer a philosophical curiosity—it's a technological resource.

Quantum computing, born from the marriage of quantum theory and information science, represents a leap so profound it's difficult to overstate. In a world built on bits—zeroes and ones—quantum computers offer a new language: one where particles can be both, entangled across space, and manipulated with surgical precision to solve problems that would take classical machines eons.

But we're not just dreaming. We're building. Qubits are no longer confined to theory—they exist in dilution refrigerators, laser traps, and photonic circuits. Algorithms once confined to blackboards are running on noisy intermediate-scale quantum (NISQ) devices in labs and data centers around the world.

This book is a journey into **advanced quantum computing**—a field that sits at the bleeding edge of science and engineering. We'll explore how quantum mechanics is harnessed for computation, how new architectures are built atom by atom, and how revolutionary algorithms promise breakthroughs in chemistry, optimization, machine learning, and cryptography.

Yet, this journey is also about uncertainty. The path forward is neither straight nor smooth. Errors, decoherence, and hardware limitations still haunt every milestone. But the challenge is also the opportunity. Those who engage now—scientists, engineers, coders, dreamers—are not just learning a new tool. They are helping define the next era of computation.

If the classical computer brought us the internet, mobile devices, and artificial intelligence, **the quantum computer may open doors we've yet to imagine**—from simulating the origin of the universe to unlocking new forms of secure communication.

This is no longer science fiction. It's science in progress.

Welcome to the next frontier.

Section 1: Foundations of Quantum Mechanics

Chapter 1

The Quantum World: A New Frontier

For centuries, the world of physics was governed by the classical laws of Newton and Maxwell—frameworks that explained the motion of celestial bodies, the workings of machines, and the nature of electricity and magnetism with impressive accuracy.

Yet, as the 20th century emerged, scientists began encountering phenomena that didn't fit these classical models. Behaviors observed at the atomic and subatomic levels appeared fundamentally different, defying what was long accepted as scientific truth.

This marked the dawn of quantum mechanics—a revolutionary field that introduced a new way of understanding nature. Unlike classical physics, which is deterministic and intuitive, quantum mechanics is probabilistic, counterintuitive, and profoundly strange.

At the heart of quantum theory lies the concept of **superposition**, which allows a particle to exist in multiple states simultaneously. Only when a measurement is made does the particle "choose" a definite state.

The famous thought experiment of Schrödinger's cat illustrates this idea. A cat placed in a box with a quantum-triggered poison is simultaneously alive and dead—until someone opens the box. In the quantum realm, such superpositions aren't just metaphors—they're real.

Equally fascinating is **entanglement**, a phenomenon where two or more particles become linked so that the state of one instantly affects the state of the other, regardless of the distance between them. Einstein called it "spooky action at a distance," and though skeptical, he was ultimately proven wrong by experiments that confirmed entanglement's reality.

While these phenomena might sound theoretical or abstract, they have already reshaped our world. Technologies such as lasers, MRI scanners, GPS systems, and semiconductors are all rooted in quantum principles.

Now, we stand on the edge of the next quantum revolution—one that takes the power of quantum mechanics and applies it to information processing. This is the promise of **quantum computing**.

Unlike classical bits, which are either 0 or 1, **quantum bits** (or qubits) can be both at once due to superposition. This allows quantum computers to evaluate many possibilities in parallel, unlocking new frontiers in speed and problem-solving.

Quantum computers also leverage entanglement to create complex correlations between qubits, enabling computational feats that classical systems simply cannot achieve.

This quantum parallelism means quantum computers are well-suited to tackle problems that are practically impossible for traditional machines—like simulating molecules, optimizing supply chains, or factoring extremely large numbers.

But quantum computing doesn't just require faster chips—it demands a completely new way of thinking. The architecture, physics, and logic behind quantum systems are fundamentally different from the classical computers we use today.

Qubits are delicate. They are highly sensitive to environmental noise and can easily lose their quantum properties—a process known as **decoherence**. To prevent this, quantum computers must be maintained at temperatures colder than outer space.

Several competing qubit technologies are being developed, including superconducting circuits, trapped ions, topological qubits, and photonic systems. Each has unique strengths and challenges, and no consensus yet exists on which platform will ultimately dominate.

The current stage of quantum development is known as the **NISQ era**—short for **Noisy Intermediate-Scale Quantum**. These devices are powerful enough to perform basic quantum tasks, but still far from being universally reliable or scalable.

Despite their limitations, NISQ devices are already enabling researchers to test early quantum algorithms and explore promising real-world applications in fields such as finance, material science, and machine learning.

The goal is to reach **quantum advantage**, the point at which a quantum computer outperforms the best classical supercomputers at a practical task. In 2019, Google claimed such an achievement—though the specific use case was narrow, it marked a pivotal milestone.

Still, a formidable challenge remains: quantum systems are inherently noisy, and even small errors can cascade and corrupt results. That's where **quantum error correction** comes in.

Through the use of sophisticated error-correcting codes, scientists can protect quantum information without destroying it—a delicate balance essential for long-term scalability.

Ultimately, **fault-tolerant quantum computing** is the holy grail of the field. Achieving it requires creating stable "logical qubits" from thousands of physical qubits, a monumental engineering and computational effort.

Yet progress continues. Tech giants like IBM, Google, and Microsoft, as well as agile startups like IonQ and PsiQuantum, are racing to build better hardware and more effective quantum software.

Governments, too, have recognized the strategic importance of quantum computing, investing billions in national research initiatives and technological infrastructure.

Unlike other computing trends, quantum computing doesn't merely improve what we already do—it changes what is possible. It redefines the boundaries of computation.

For businesses, quantum computing presents a dual narrative: it may render current encryption obsolete, but it also opens the door to entirely new competitive advantages in modeling, logistics, and forecasting.

For scientists, it offers unprecedented tools to simulate complex molecular structures, potentially revolutionizing medicine, climate science, and materials engineering.

For society, quantum raises profound ethical and geopolitical questions. Who controls the technology? How is it regulated? How do we prepare our workforce?

Educational institutions are beginning to integrate quantum topics into curricula, while cloud-based platforms like IBM Quantum and Amazon Braket are making quantum tools accessible to developers and students around the world.

Still, mastering quantum computing requires interdisciplinary expertise—in physics, computer science, mathematics, and electrical engineering—making collaboration more vital than ever.

As with the early days of classical computing, we can expect setbacks, hype cycles, and unexpected discoveries. But the long-term trajectory is unmistakable: quantum computing will be transformative.

The potential impact spans from pharmaceuticals to national defense, from clean energy solutions to AI optimization. What once seemed speculative now inches closer to engineering reality.

Just as classical computing evolved from bulky mainframes to powerful smartphones, quantum computing will take time to mature—but its foundational importance is already clear.

The quantum world is no longer confined to textbooks or particle accelerators. It's becoming programmable, tangible, and accessible.

This is the new frontier: a convergence of physics and computation, of science fiction and fact, of mystery and mastery.

And while the journey ahead is complex, it is also rich with promise. Quantum computing doesn't just offer more speed or efficiency—it offers new answers to old questions.

In the chapters that follow, we will navigate this emerging domain, exploring its science, technologies, and real-world applications. This is not just a story of qubits—it's a story of what comes next.

Welcome to the quantum age.

Chapter 2

Superposition - The Power of Possibilities

In the classical world, a light switch is either on or off. A coin is either heads or tails. A bit in a computer is either a 0 or a 1. Our intuition is shaped by these discrete, binary choices—states that exist independently and exclusively.

But in the quantum world, these rules do not apply.

Here, reality is governed by probabilities and possibilities, not absolutes. At the heart of this paradigm shift lies one of the most profound principles in physics: **superposition**.

Superposition is the ability of a quantum system to exist in multiple states simultaneously. A quantum bit—or **qubit**—can be in a state of 0, a state of 1, or any **complex linear combination** of both. It does not "choose" a definitive state until it is observed or measured.

This concept challenges everything we know about how the world works. Imagine a light switch not just being halfway between on and off, but being both fully on and fully off at the same time—until someone looks at it. That's the bizarre but real nature of superposition.

To represent this mathematically, consider the basic states of a qubit:
$|0\rangle$ and $|1\rangle$. A general quantum state can be written as:

$\psi = \alpha|0\rangle + \beta|1\rangle$,
where α and β are complex numbers, and the sum of their squared magnitudes is 1 (i.e., $|\alpha|^2 + |\beta|^2 = 1$).

These coefficients represent the **probability amplitudes** of the system collapsing into either state upon measurement. They encode both magnitude and phase, offering richer information than classical probabilities.

When a measurement is made, the system "collapses" into one of its basis states, with the probabilities determined by α and β. Until that moment, however, the system exists in a delicate, coherent superposition of both.

This isn't just theoretical. Superposition is a **measurable, testable** feature of quantum mechanics. It has been confirmed in countless experiments, from electrons in double-slit tests to superconducting circuits in modern quantum computers.

In quantum computing, superposition is not just a curiosity—it's a **computational resource**. It enables a qubit to process many possibilities at once. A system of **n qubits** can represent $\mathbf{2^n}$ states simultaneously.

For instance, 3 classical bits can store just one of eight possible combinations at a time. But 3 qubits in superposition can represent all 8 combinations at once. As the number of qubits increases, this exponential scaling leads to a powerful increase in parallelism.

It's tempting to think of this as "quantum computers doing all computations at once," but that's not quite accurate. The power of superposition is not brute-force parallelism—it's the ability to manipulate interference between quantum states to amplify correct outcomes and cancel out incorrect ones.

That's what quantum algorithms like **Grover's** and **Shor's** exploit: not just superposition, but interference and entanglement as well. These quantum features combine to allow efficient exploration of vast solution spaces that would be infeasible classically.

But superposition is fragile. Quantum states are highly susceptible to **decoherence**, the loss of quantum information due to environmental interaction. Even the smallest disturbance—heat, electromagnetic noise, cosmic rays—can destroy the delicate balance.

This is why quantum computers must operate at temperatures near absolute zero and be isolated from even the faintest disturbances. Maintaining superposition over time is one of the grand engineering challenges of quantum computing.

Yet, the more we learn to stabilize and control quantum states, the more we unlock their power. The **quantum logic gates** used to manipulate qubits—such as the Hadamard gate—are designed specifically to induce and manage superpositions.

For example, applying a Hadamard gate to a qubit in the state $|0\rangle$ transforms it into an equal superposition:
$\mathbf{H|0\rangle = (1/\sqrt{2})(|0\rangle + |1\rangle)}$.
Now, the qubit has a 50/50 chance of being measured as either 0 or 1, but exists in both states until that measurement.

This ability to encode information probabilistically opens doors to new algorithmic strategies. Instead of trying every answer one by one, quantum systems can explore many pathways at once, guided by quantum interference.

Superposition also underlies quantum parallelism in **quantum simulation**, allowing researchers to model the behavior of molecules, atoms, and materials with unprecedented precision.

In practical terms, this could lead to breakthroughs in **drug discovery**, **chemical reactions**, **battery design**, and **new materials**—all areas where classical simulations struggle with the complexity of quantum interactions.

It's important to note that while superposition offers enormous promise, it's not a magic bullet. Quantum algorithms must be designed carefully to take advantage of it. Most problems are not naturally suited for quantum speed-up, and many classical tasks remain best handled by traditional systems.

Nonetheless, the strategic value of superposition cannot be overstated. It represents a profound shift in our approach to computation—from deterministic logic to probabilistic exploration, from rigid control to dynamic orchestration of possibility.

We are still in the early days of learning how to program and harness superposition. As researchers develop new error correction techniques, more stable qubit designs, and robust algorithms, the full power of this quantum principle will come into focus.

In a sense, superposition mirrors the human experience. Before we make choices, we live in a space of possibilities. The future is undefined until acted upon. In the quantum world, this ambiguity is not a limitation—it's a strength.

It challenges us to rethink what it means to "know" something, to compute something, and to understand the very nature of information.

As we build systems that can inhabit and manipulate these quantum possibilities, we aren't just building faster machines—we're building entirely new ways of interacting with the fabric of reality.

Superposition is the foundation of quantum computing. It is not just a state—it is a **strategy**, a **framework**, and a **philosophy** of computation.

In the coming chapters, we will see how this principle intertwines with other quantum phenomena—such as entanglement and interference—to create the astonishing potential of quantum information processing.

But it all begins here: with the power of possibilities, encoded in every quantum bit, waiting to be measured, shaped, and transformed into the technology of tomorrow.

Chapter 3

Entanglement - The Spooky Connection

In the quantum world, distance doesn't always mean separation. Two particles, born together or brought into interaction, can become linked in such a profound way that their fates are forever intertwined. This phenomenon is called **entanglement**, and it lies at the very heart of quantum mechanics—and the future of quantum computing.

To the classical mind, entanglement sounds impossible. If you separate two coins, you'd expect them to behave independently. Flip one, and its result says nothing about the other. But quantum particles don't play by these rules. In an entangled state, the measurement of one particle instantly determines the state of its partner, no matter how far apart they are.

This "instantaneous" effect is what troubled Albert Einstein, who famously referred to it as **"spooky action at a distance."** He doubted it could be real, suspecting some hidden variables might explain the correlations. But experiments have since shown—repeatedly and decisively—that quantum entanglement is real, measurable, and not the result of any hidden classical mechanism.

Entanglement begins when two qubits interact in a way that causes their quantum states to become interdependent. Once entangled, any operation or measurement on one qubit affects the state of the other, even if they are physically separated by kilometers.

To illustrate, imagine two qubits in the entangled state:
$(1/\sqrt{2})(|00\rangle + |11\rangle)$
This means that if one qubit is measured and found to be in state $|0\rangle$, the other will **immediately** be found in state $|0\rangle$. Likewise, if one is $|1\rangle$, the other is guaranteed to be $|1\rangle$—regardless of how far apart they are.

Crucially, these outcomes are not predetermined. They exist in a superposition until one is observed. It's the measurement of one qubit that forces both into a correlated state—what physicists call "wave function collapse."

Entanglement defies locality, one of the bedrock assumptions of classical physics—that information can only travel at or below the speed of light. Entangled particles seem to "communicate" faster than light, though in truth, no usable information is transmitted faster than light, preserving the broader framework of causality.

So how does this relate to quantum computing?

Entanglement allows **quantum systems to scale exponentially**, enabling multiple qubits to act as a single, unified computational object. Classical bits operate independently. Quantum bits, through entanglement, operate interdependently, creating a tightly coupled network of information.

This coupling is what gives quantum computers their **unprecedented processing potential**. Entangled qubits can represent and manipulate large state spaces more efficiently than classical systems ever could.

Moreover, entanglement enables **quantum logic gates** like the **CNOT gate**, which flips one qubit depending on the state of another. These multi-qubit operations are the building blocks of quantum algorithms and impossible to replicate classically.

Entanglement also plays a vital role in **quantum teleportation**, a technique in which the state of a qubit is transferred from one location to another—without physically moving the qubit itself.

While it doesn't involve teleporting matter, this process transfers **quantum information** securely and instantaneously across distance, enabled entirely by entanglement.

Beyond computing, entanglement underpins **quantum cryptography**, particularly **Quantum Key Distribution (QKD)**protocols like BB84 and E91. These systems offer theoretically unbreakable encryption because any eavesdropping attempt breaks the entanglement and reveals the intrusion.

Quantum entanglement also shows up in experimental research on **quantum networks**, where multiple quantum devices are linked via entangled photons to share quantum information. This is the foundational concept behind the future **quantum internet**—a global-scale network for secure quantum communication and distributed quantum computing.

In a deeper sense, entanglement challenges the very notion of individuality in the quantum realm. When particles are entangled, they lose their distinct identities. Their combined state becomes more fundamental than the parts—an idea that continues to reshape the way we understand information, identity, and causality.

Yet entanglement is also fragile. Like superposition, it is easily destroyed by **decoherence**—interactions with the environment that collapse the delicate quantum state. That's why entangled systems must be carefully shielded and cooled, often requiring cryogenic hardware to remain stable.

Preserving entanglement over time and distance is one of the greatest technical challenges in building practical quantum computers. Engineers must design systems that not only create entanglement reliably but also **maintain it** throughout the computation and measurement process.

To combat decoherence, researchers are developing **quantum error correction codes** that rely heavily on entanglement. These codes distribute quantum information across multiple entangled qubits, so that if some are lost or corrupted, the original information can still be reconstructed.

This concept of **redundancy through entanglement** mirrors the way classical systems use parity bits or RAID storage—but with vastly more complexity and mathematical nuance.

In practice, the most powerful quantum algorithms—including **Shor's algorithm** for factoring and **quantum simulation techniques**—derive their computational advantage from entangling multiple qubits and manipulating those entangled states with precision.

Thus, entanglement is not just a curiosity. It is **an essential ingredient** in almost every aspect of quantum computing: hardware design, algorithmic power, error correction, communication, and long-term scalability.

Einstein may have called it spooky, but today entanglement is no longer a mystery—it's a tool, one that scientists and engineers are learning to wield with increasing skill.

And the implications extend far beyond computing. Entanglement has triggered discussions in philosophy, metaphysics, and even consciousness studies, as we grapple with its nonlocal, probabilistic nature.

It forces us to reconsider the boundaries between systems, the meaning of measurement, and the role of the observer in shaping reality.

As we build more complex quantum devices, our ability to control and exploit entanglement will define the trajectory of progress. Whether we're simulating the behavior of new drugs or transmitting quantum keys across continents, entanglement will be at the core.

In the coming chapters, we'll explore how entanglement is harnessed in real-world quantum hardware, how it's implemented in algorithms, and how it enables concepts like quantum teleportation and entanglement swapping.

But for now, let us pause to appreciate what entanglement truly represents: a bridge between particles, a union of possibilities, and a doorway into the most connected, mysterious, and beautiful aspects of the quantum universe.

Chapter 4

Quantum Interference and Measurement

Quantum mechanics is full of paradoxes, but few are as central—and as useful—as the phenomenon of **quantum interference**. It's the principle that allows quantum computers to extract meaningful answers from a sea of possibilities. But this power only becomes useful when paired with its curious counterpart: **measurement**.

To understand quantum interference, we must return to the idea of **superposition**. A qubit can exist in a combination of states, each with a certain probability amplitude. These amplitudes are complex numbers that carry both magnitude and **phase**—a key element in interference.

In classical probability, events add linearly. But in quantum systems, **amplitudes** add, and then the probability is calculated by squaring the **resulting total amplitude**. This opens the door to interference—constructive or destructive—based on how the phases align.

The most famous demonstration of quantum interference is the **double-slit experiment**. When electrons or photons pass through two slits without being observed, they don't behave like particles—they form an interference pattern on a screen, just like waves would. Peaks and troughs reinforce or cancel each other out, producing bright and dark regions.

But when one attempts to observe which slit the particle goes through, the interference pattern disappears. The act of measurement **collapses** the wave function, forcing the system to commit to a specific path and destroying the delicate balance of interference.

In quantum computing, this dual behavior of quantum states—interfering when unobserved and collapsing when measured—is harnessed deliberately to design algorithms.

Quantum interference allows a quantum system to amplify correct solutions while suppressing incorrect ones. This is not random—it's orchestrated through **quantum gates** that rotate qubit states in such a way that favorable computational paths interfere constructively, while others cancel out.

This idea is central to **Grover's algorithm**, which searches an unsorted database of N items in roughly $\sqrt{N}$ steps—quadratically faster than any classical method. The algorithm uses repeated interference patterns to increase the probability amplitude of the correct answer.

Measurement in quantum mechanics is unlike anything in classical computing. When a qubit is measured, it no longer exists in superposition. It collapses into either $|0\rangle$ or $|1\rangle$, with probabilities determined by the square of the amplitude of each state.

This is why measurement must be delayed until the end of a quantum computation. Measuring too early collapses the state prematurely, potentially discarding useful computational potential before it's fully exploited.

Moreover, because measurement yields a **probabilistic outcome**, quantum computations are typically repeated multiple times to collect statistical data. The correct result appears more frequently if the algorithm has been well-designed to maximize its amplitude.

But measurement isn't just about outcomes—it's also a tool for **feedback** and **error correction**. Quantum error-correcting codes often rely on measuring **ancilla qubits** (helper qubits) to detect faults in the system without disturbing the logical quantum information.

This subtle form of measurement—called **indirect measurement** or **syndrome extraction**—allows a quantum computer to diagnose errors without collapsing the valuable state it is trying to protect.

In this way, measurement becomes a **strategic component** of quantum computing. It is both the end point and the safeguard, used to extract information and preserve integrity.

However, measurement in quantum mechanics remains one of the most philosophically rich and debated concepts. What exactly constitutes a "measurement"? Does it require a conscious observer, or is interaction with any macroscopic system enough?

These questions, while speculative, remind us that quantum computing isn't just a technical field—it's a direct engagement with the deepest foundations of reality.

From a practical standpoint, **quantum interference** is what gives quantum algorithms their power. Superposition lets quantum systems explore many possibilities, but interference is what guides them to the correct solution.

Without interference, a quantum computer would merely produce a noisy mix of probabilities. Without measurement, we'd never extract anything from that mix.

Together, interference and measurement form the dynamic engine of quantum computing—one generating a space of rich possibilities, the other collapsing it into meaningful, actionable outcomes.

Engineers and physicists have learned to manipulate interference through gates like Hadamard, phase, and controlled gates. These are choreographed in circuits that sculpt amplitude distributions across qubit states.

As quantum computers grow in size and complexity, interference patterns become harder to predict—but also more powerful to exploit. Understanding and optimizing interference is a key frontier in quantum algorithm research.

Similarly, quantum measurement systems must become more precise, faster, and more robust. Innovations in **quantum sensors**, **readout electronics**, and **error detection protocols** are pushing the boundaries of how much we can see without collapsing too much.

In the years ahead, researchers aim to control measurement so precisely that they can **steer** quantum systems mid-computation without fully collapsing their wave functions—a technique called **quantum feedback control**.

This would allow for dynamic, adaptive quantum computing systems—machines that adjust their computational paths in real time, based on partial observations, rather than fixed instruction sets.

Ultimately, quantum interference and measurement reveal the delicate dance at the heart of quantum computation: between exploration and precision, between possibility and decision, between openness and collapse.

It's a reminder that the quantum world is not about certainty, but **probabilistic design**. A world where solutions are not calculated directly, but rather sculpted into view by orchestrating subtle patterns of constructive and destructive interference.

In this strange and powerful space, measurement doesn't just reveal answers—it **selects** them. Interference doesn't just mix probabilities—it **designs** the outcome landscape.

Together, they enable the leap from theoretical possibility to practical computation. And as we move forward into the realm of full-scale quantum computing, mastering this interplay will be essential—not just for building faster machines, but for understanding the very nature of information itself.

Chapter 5

The No-Cloning Theorem and Quantum Information

In the classical world, copying information is easy. We duplicate files, back up drives, and replicate data across servers without a second thought. But in the quantum world, the rules are different—strikingly different.

Here, a central law of nature tells us that **perfectly copying an arbitrary unknown quantum state is impossible**. This rule is known as the **No-Cloning Theorem**, and it is one of the most surprising—and essential—features of quantum mechanics.

First formulated in the early 1980s by Wootters, Zurek, and Dieks, the No-Cloning Theorem states that **there is no physical process that can duplicate an unknown quantum state without disturbing it**. If you don't know the exact state of a qubit, you cannot make a perfect copy.

This is not a technological limitation—it's a **fundamental law of quantum physics**, rooted in the linearity and unitarity of quantum mechanics. It is as unbreakable as the speed of light or the conservation of energy.

To understand why this matters, let's contrast with the classical case. If you have a bit in state 0 or 1, copying it is trivial. It can be read, replicated, and transmitted as many times as needed. But if you have a qubit in an arbitrary superposition state, say:

$\psi = \alpha|0\rangle + \beta|1\rangle$,

you cannot build a device that takes ψ and an empty state $|0\rangle$, and outputs $\psi \otimes \psi$ for arbitrary α and β.

Any attempt to copy the qubit disturbs the original state or produces a fundamentally incorrect result—unless the state is already known, in which case you're not cloning, you're re-preparing.

This unique behavior has deep implications.

First, it safeguards **quantum information** in a way classical information is not. Because quantum states cannot be copied, they cannot be intercepted and reproduced undetectably. This is the principle behind **quantum cryptography**, where the act of eavesdropping changes the information being transmitted and reveals the intrusion.

In **Quantum Key Distribution (QKD)** protocols like BB84, the No-Cloning Theorem is what makes it impossible for an adversary to intercept a quantum key and copy it without detection. Any measurement attempt collapses the qubit's superposition, altering the state and exposing the breach.

Second, the No-Cloning Theorem informs the design of **quantum error correction**. In classical computing, redundancy is achieved by copying data. In quantum systems, that's not an option. Instead, **entangled states** and **syndrome measurements** are used to protect quantum information without actually copying it.

For example, the **Shor code** and **surface code** encode a single logical qubit into a large entangled state of many physical qubits, allowing for errors to be detected and corrected without directly measuring or duplicating the original quantum information.

Interestingly, while cloning is forbidden, **teleportation is not**. **Quantum teleportation** allows the state of a qubit to be transferred from one location to another—using entanglement and classical communication—without physically moving the particle or cloning it.

The catch is that the original state must be **destroyed** in the process. The no-cloning law is preserved because teleportation doesn't create a copy; it **relocates** quantum information.

This may seem like a constraint, but it is actually a feature that ensures **quantum coherence** and **security** in the manipulation and transfer of information. It forces us to design protocols that are more subtle, more secure, and more elegant.

Another consequence of the No-Cloning Theorem is that **quantum information is private by nature**. There is no way to look inside a quantum system and fully extract its state without fundamentally disturbing it. This property distinguishes **quantum information theory** from its classical counterpart.

It also challenges our intuitions about knowledge and measurement. In classical physics, knowing the state of a system allows us to duplicate and predict its future. In quantum physics, knowledge is contextual and limited—**you can never know everything** about a quantum system, and you can never copy what you don't fully know.

This isn't just a curiosity—it is the cornerstone of **quantum information theory**, a field that merges quantum mechanics with information science to explore how information is stored, transmitted, and processed in a quantum universe.

The No-Cloning Theorem also forces us to rethink **quantum networking**. In a classical network, messages can be cached, copied, and relayed without concern. In a quantum network, each bit of quantum data (qubit) must be handled with care, ensuring it's neither measured nor cloned inadvertently.

This leads to the development of **quantum repeaters**—devices that extend quantum communication over long distances without copying. They use entanglement swapping and purification techniques, rather than amplification and duplication, to maintain integrity over vast networks.

In quantum computing, this no-cloning rule complicates **checkpointing**, **state inspection**, and **debugging**. Developers must build systems that extract insight without collapsing or destroying the computation, using **indirect measurements**, **ancilla qubits**, and **quantum tomography** techniques.

Yet despite these challenges, the No-Cloning Theorem is a **blessing in disguise**. It's what enables **unforgeable quantum money**, **secure authentication protocols**, and **proofs of identity** that are impossible to counterfeit in principle.

It represents a radical shift from the world of classical redundancy to a world where **uniqueness is power**, where the inability to copy protects value, and where information behaves less like a static commodity and more like a delicate state of being.

In summary, the No-Cloning Theorem is not just a limit—it's a **pillar** of the quantum world. It defines how we interact with quantum information, how we secure it, and how we move it from place to place.

It shapes the architecture of quantum computers, the structure of quantum communication, and the philosophy of quantum information science.

As we build the quantum technologies of tomorrow, we must remember that in this new world, duplication is destruction, observation is disruption, and fidelity is earned not by copying, but by coherence and control.

The No-Cloning Theorem reminds us that quantum information is not just data—it is **contextual, dynamic, and uniquely fragile**. And understanding it is essential to mastering the next generation of computation, security, and communication.

Chapter 6

Qubits vs. Classical Bits

Every computing revolution begins with a new unit of information.

In classical computing, that unit is the **bit**—the simplest building block of digital logic. A bit is binary. It can be either 0 or 1. No in-between. This simplicity is what powers everything from smartphones to supercomputers, built on transistors that switch between these two stable states.

In quantum computing, the fundamental unit of information is the **qubit**—short for **quantum bit**. At first glance, it may seem similar to a classical bit. It also has two basis states, $|0\rangle$ and $|1\rangle$. But unlike a bit, a qubit can exist in a **superposition** of both states at once.

This is the first and most profound difference: **a classical bit stores a single definite value at any time, while a qubit stores a range of possibilities** until measured.

Mathematically, a qubit's state is represented as:
$|\psi\rangle = \alpha|0\rangle + \beta|1\rangle$,
where α and β are complex numbers that define the **probability amplitudes** of each state. The probabilities themselves are given by $|\alpha|^2$ and $|\beta|^2$, and must always sum to 1.

The presence of both real and imaginary components in a qubit's state allows for **quantum phase**, which is invisible in classical systems but essential in quantum computation. Phase enables **interference**, where quantum states can reinforce or cancel each other—giving quantum algorithms their edge.

A classical system with n bits can represent **one of 2^n** possible states at a time. A quantum system with n qubits, in contrast, can represent **a superposition of all 2^n states simultaneously**. This is

quantum parallelism—the ability to encode and process massive amounts of information in a compact form.

However, there's a caveat: even though a quantum system can exist in a superposition of many states, **measuring it only yields one outcome**. This is why quantum algorithms must be designed carefully to **manipulate the amplitudes** so that the correct answers become more likely upon measurement.

In classical computing, copying data is simple and ubiquitous. Bits can be duplicated, stored, transmitted, and backed up at will. In contrast, quantum information **cannot be copied** due to the **No-Cloning Theorem**, which states that it is impossible to create an identical copy of an arbitrary unknown quantum state.

This makes qubits fundamentally more fragile than bits. Any interaction with the environment can cause **decoherence**, collapsing the quantum state and destroying the information. As a result, qubits must be carefully shielded, cooled, and controlled to preserve their quantum properties.

Another major difference lies in **logic operations**. Classical computers use logic gates like AND, OR, and NOT to perform operations on bits. These gates are irreversible—a NOT gate flips a bit, but once flipped, you can't recover the original state without extra information.

Quantum gates, on the other hand, are **reversible and unitary**. A quantum gate must preserve the total probability across all possible states. Gates like **Hadamard**, **Pauli-X**, **CNOT**, and **Toffoli** manipulate qubit states in ways that are **coherent**—meaning they preserve the superpositions and phases of the system.

The **Hadamard gate**, for instance, creates a superposition. The **CNOT gate** entangles two qubits, introducing correlations that classical gates cannot replicate. The complex choreography of quantum gates is what enables powerful algorithms like Shor's and Grover's.

Qubits also differ in **entanglement**. In classical systems, bits are independent. The state of one bit doesn't affect another. In quantum systems, qubits can become entangled, meaning the state of one qubit is intrinsically linked to the state of another, even across great distances. This entanglement enables distributed computation, quantum teleportation, and the exponential speedup in certain problems.

Yet another contrast is in **error correction**. Classical bits use redundancy—copying bits multiple times to detect and correct errors. Since qubits can't be cloned, quantum error correction relies on **entanglement and syndrome measurements** to detect and fix errors without collapsing the quantum state.

Physically, bits are stored in stable voltage states in transistors. Qubits, by contrast, are realized in a variety of physical systems—superconducting loops, trapped ions, photons, quantum dots—each with its own trade-offs in speed, coherence time, and scalability.

From an information theory perspective, bits obey **Shannon entropy**, which defines the limits of classical communication and compression. Qubits, on the other hand, fall under **von Neumann entropy**, and their behavior opens the door to new limits in secure communication and data processing.

Importantly, bits are deterministic and predictable. Given the initial state and input, a classical computation will always yield the same result. Quantum computing is **probabilistic by nature**, and outputs must often be interpreted statistically—based on repeated measurements across many runs.

This stochastic character can be unsettling at first, but it also offers advantages. Quantum systems can **explore complex solution landscapes** more naturally, making them ideal for optimization, simulation, and machine learning tasks that involve ambiguity and uncertainty.

The differences between bits and qubits represent not just a technological shift but a **paradigm shift**. Qubits are not upgrades to bits—they are a new form of information that behave according to entirely different rules.

They challenge classical assumptions about duplication, determinism, and locality. They force us to design new hardware, write new languages, and create entirely new algorithmic strategies.

In practical terms, qubits are still far from replacing bits in general-purpose computing. Classical computers remain vastly more reliable and scalable—for now. But in domains where quantum effects can be harnessed effectively, qubits offer solutions that are fundamentally beyond the reach of any classical system.

As the field of quantum computing matures, the hybrid use of bits and qubits—**quantum-classical integration**—will become increasingly important. Systems will need to translate between the deterministic logic of bits and the probabilistic amplitudes of qubits.

To understand qubits is to step into a new worldview—where uncertainty is power, where parallelism is native, and where information behaves more like a fluid wave than a rigid structure.

The bit brought us the digital revolution. The qubit may bring us the next one.

Section 2: Building Quantum Computers

Chapter 7

Quantum Logic Gates and Circuits

In classical computing, logic gates are the fundamental components of digital circuits. AND, OR, NOT, NAND, NOR—these basic operations manipulate bits and form the basis for all computational logic. They are irreversible, binary, and deterministic.

Quantum computing, too, has its own version of logic gates, but they operate under profoundly different rules. **Quantum logic gates** are the essential operations that manipulate **qubits**, and they are governed by the principles of **quantum mechanics**, not classical Boolean algebra.

The most important distinction is this: **quantum gates are reversible**. Each gate corresponds to a **unitary transformation**—a mathematical operation that preserves the total probability (or norm) of a quantum state. This reversibility is not just a design preference; it's a **physical necessity** arising from the linear and unitary nature of quantum evolution.

A quantum logic gate acts on one or more qubits and transforms them from one quantum state to another. These gates form the building blocks of **quantum circuits**, much like how classical logic gates form classical circuits. But instead of manipulating fixed values of 0 or 1, quantum gates manipulate **probability amplitudes**, **phases**, and **entanglement**.

Let's begin with the simplest case: **single-qubit gates**. These operate on individual qubits and are represented by 2×2 unitary matrices.

The most fundamental single-qubit gate is the **Pauli-X gate**, also called the quantum NOT gate. It flips the state of a qubit:

- $X|0\rangle = |1\rangle$
- $X|1\rangle = |0\rangle$

Next, the **Hadamard gate (H)** plays a critical role by putting qubits into a superposition. Applying H to $|0\rangle$ yields:
$H|0\rangle = (1/\sqrt{2})(|0\rangle + |1\rangle)$
This creates an equal superposition of 0 and 1, allowing the quantum computer to explore multiple computational paths simultaneously.

The **Pauli-Z gate** flips the phase of the $|1\rangle$ component, leaving $|0\rangle$ unchanged:

- $Z|0\rangle = |0\rangle$
- $Z|1\rangle = -|1\rangle$

There are also **rotation gates**, such as **Rx(θ)**, **Ry(θ)**, and **Rz(θ)**, which rotate the qubit state around different axes of the Bloch sphere by angle θ. These provide fine-grained control over the quantum state, essential for quantum algorithms and variational circuits.

Now let's explore **multi-qubit gates**, which enable interaction between qubits—crucial for building entanglement.

The most well-known is the **CNOT (Controlled-NOT) gate**. It flips the target qubit if the control qubit is $|1\rangle$, and does nothing if the control is $|0\rangle$.
CNOT is the quantum analog of XOR, and more importantly, it is the **foundation of entanglement**.

Consider this simple entangling operation:

- Start with $|00\rangle$
- Apply a Hadamard gate to the first qubit:
  $\rightarrow |\psi\rangle = (1/\sqrt{2})(|0\rangle + |1\rangle)|0\rangle$

- Apply a CNOT with the first qubit as control and the second as target:
  → Result: (1/√2)(|00⟩ + |11⟩), a **Bell state**—a maximally entangled pair of qubits.

Other important two-qubit gates include:

- **Controlled-Z (CZ)**: Flips the phase of the target qubit if the control is |1⟩.
- **SWAP gate**: Exchanges the states of two qubits.
- **Toffoli gate (CCNOT)**: A three-qubit gate that flips the target if both controls are |1⟩. This gate is universal for classical reversible computing and plays a key role in error correction.

Unlike classical gates, quantum gates don't just apply logical transformations—they **preserve coherence** and **interference patterns** across the quantum system.

Quantum circuits are sequences of these gates applied to qubits over time. The circuit is designed so that when the qubits are measured at the end, the interference effects will **enhance** the correct solution's probability and **suppress** the incorrect ones.

Quantum circuit design is both an art and a science. Every gate added introduces **noise and error**, especially in current NISQ-era devices. So circuits must be optimized to use as few gates as possible while achieving the desired quantum transformation.

Most quantum programs use **circuit diagrams**, where time flows from left to right, horizontal lines represent qubits, and gates are applied as symbols along those lines. This visual format provides a clear way to see how states evolve step-by-step.

Quantum circuits can also be described in code using frameworks like **Qiskit** (IBM), **Cirq** (Google), and **Pennylane**(Xanadu). These libraries translate high-level quantum instructions into hardware-executable gate sequences.

Another key concept in quantum circuits is **universality**. A set of quantum gates is said to be universal if any quantum operation (unitary transformation) can be approximated to arbitrary accuracy using only gates from that set. For instance, the set {Hadamard, CNOT, T} is universal.

Quantum circuits are not just theoretical constructs—they are physical processes. In superconducting systems, gates are implemented using microwave pulses. In trapped ion systems, laser pulses manipulate the quantum states. Each implementation has its own speed, error rates, and connectivity constraints.

Because of the limitations of today's hardware, much of quantum circuit design focuses on **depth minimization**(shorter sequences), **gate fidelity**, and **qubit connectivity mapping**—which qubits can directly interact with which others.

One of the challenges in scaling quantum computers is how to manage gate **errors**. Every gate has a small but non-zero chance of failing. As circuits grow, so does the accumulated error. This makes **fault-tolerant circuit design** and **error correction codes** essential for long-term progress.

In practical quantum algorithms like Shor's (for factoring) and Grover's (for search), the structure of the quantum circuit determines the speedup and correctness of the outcome. Even minor flaws in gate design or sequencing can destroy the computation's success probability.

Ultimately, quantum gates and circuits represent **quantum computation in action**. They allow us to transform abstract mathematical operations into real physical processes, executed by quantum hardware at near-absolute zero temperatures.

Where classical logic circuits build answers bit-by-bit, quantum logic circuits build **probability landscapes**, subtly guiding the system toward high-probability correct answers.

As we continue to develop more qubits, better coherence, and improved gate fidelity, quantum circuits will become more complex, more reliable, and more powerful.

In the next chapters, we'll explore how these circuits are used to build real algorithms and simulate real-world systems—from chemistry and physics to machine learning and optimization.

But first, it's important to appreciate the elegance and constraint of the quantum logic gate: precise, reversible, and yet capable of doing the impossible—**manipulating possibility itself**.

Chapter 8

Quantum Hardware Architectures

At the heart of every quantum computer lies a delicate physical system designed to embody one of the strangest ideas in modern science: the qubit. But turning that theoretical concept into a working, scalable device requires a deep interplay between physics, engineering, and information theory.

Unlike classical computers—which are built from silicon-based transistors on integrated chips—quantum computers demand **exotic materials**, **extreme environments**, and **precise quantum control**. The quantum world is fragile, and building stable quantum hardware is one of the most technically challenging feats in science today.

There is no single way to build a quantum computer. In fact, multiple competing **quantum hardware architectures** exist, each leveraging different physical systems to create, manipulate, and measure qubits.

These platforms all aim to fulfill a common goal: maintain quantum coherence long enough to perform meaningful computations, while scaling to many qubits and minimizing errors.

1. Superconducting Qubits

Currently one of the most commercially advanced platforms, **superconducting qubits** are used by companies like IBM, Google, and Rigetti. These qubits are made from tiny loops of superconducting metal interrupted by **Josephson junctions**, which create non-linear inductance—essential for defining two quantum states.

These devices are fabricated using techniques similar to traditional microelectronics and are operated at **cryogenic temperatures** (near absolute zero) using dilution refrigerators. At these temperatures, superconductivity emerges, allowing electrons to flow without resistance and enabling stable qubit operations.

Control and readout of superconducting qubits are done using **microwave pulses**, which manipulate the quantum state and enable fast gate operations. These systems benefit from well-understood manufacturing methods and relatively fast gate times (tens of nanoseconds).

However, they are also prone to **short coherence times** (often in the range of 100 microseconds or less) and require precise error mitigation strategies. Scalability is possible, but interconnects and thermal management remain significant challenges.

2. Trapped Ion Qubits

Trapped ion systems, championed by companies like IonQ and Quantinuum, use **individual charged atoms (ions)** as qubits. These ions are suspended in electromagnetic traps using electric and magnetic fields inside a vacuum chamber.

Qubit states are encoded in the internal electronic states of the ions, and quantum gates are implemented using tightly focused **laser beams** that manipulate these states via precise quantum transitions.

Trapped ions offer exceptionally **long coherence times**, sometimes measured in seconds or even minutes—orders of magnitude longer than superconducting qubits. They also provide **high-fidelity gates** and all-to-all qubit connectivity within a single trap.

Their downside is **slower gate speed** (in the range of microseconds to milliseconds), and engineering challenges arise as the number of ions increases—due to heating, crosstalk, and complexity of controlling many optical beams.

3. Photonic Qubits

In photonic quantum computing, qubits are encoded in the quantum states of **light particles (photons)**—such as their polarization or path. Companies like Xanadu and PsiQuantum are developing systems that use **integrated photonic chips** to generate, manipulate, and measure these light-based qubits.

Photonic systems have the advantage of **room-temperature operation**, natural compatibility with optical networks, and minimal decoherence since photons don't interact with their environment easily.

However, creating strong interactions between photons for gate operations is difficult, requiring **nonlinear materials**, **interference-based logic**, or **measurement-induced gates**. Building large-scale photonic processors remains an open engineering frontier.

4. Spin Qubits and Quantum Dots

Another promising platform is **spin qubits**, where quantum information is stored in the spin state of an electron trapped in a **semiconductor quantum dot**. These systems use materials like silicon or gallium arsenide, leveraging decades of semiconductor research.

Spin qubits offer the potential for **scalability through integration** with existing chip technologies and **extremely small physical footprints**. Gate control is done using microwave or electric fields.

They show promise for low-power, scalable designs, but coherence times and error rates are still areas of active research. Intel and various university labs are actively developing this architecture.

5. Topological Qubits

Still largely theoretical and in early experimental phases, **topological qubits** aim to achieve fault tolerance through the inherent properties of **non-Abelian anyons**—exotic quasiparticles that encode information in the braiding of their world-lines.

Microsoft is one of the primary champions of this approach, attempting to build **Majorana zero modes** in hybrid nanowire-superconductor systems.

Topological qubits could, in principle, offer **intrinsic protection from noise** and decoherence, making them highly error-resilient. However, they are extremely difficult to realize experimentally, and to date, no topological qubit has been conclusively demonstrated in a scalable fashion.

6. Neutral Atom Qubits

Neutral atom systems use **ultracold atoms** trapped in arrays using **optical tweezers**—tight beams of laser light that hold and move individual atoms. Rydberg excitations are used to enable strong interactions between atoms for two-qubit gates.

These systems offer **natural scalability**—hundreds of atoms can be arranged in 1D, 2D, or even 3D geometries. Companies like QuEra are building machines based on this architecture, exploring its advantages in reconfigurability and high fidelity.

Challenges remain in laser control, atom loss, and gate speed, but progress is rapid and the technology shows great potential for mid- to large-scale quantum processors.

Comparing Architectures

Each hardware platform represents a different **trade-off triangle** among:

- **Coherence time**

- **Gate fidelity**
- **Scalability**
- **Speed**
- **Engineering complexity**

For example:

- **Superconducting qubits** are fast and manufacturable, but have short coherence times.
- **Trapped ions** have great coherence but are slower and harder to scale.
- **Photonic systems** are mobile and stable, but complex to entangle and interact.

There's no clear winner yet. Many experts believe the future will involve **hybrid systems**, where different qubit types serve different functions, and classical systems handle orchestration and error tracking.

The Road Ahead

As the quantum race heats up, hardware innovation will determine which platforms achieve **quantum advantage** in real-world applications. The next-generation machines must address:

- **Error correction scalability**
- **Cryogenic infrastructure**
- **Control electronics integration**
- **Interconnect strategies for modular quantum computing**

Emerging approaches like **modular architectures**, **3D qubit layouts**, and **cryogenic CMOS** controllers are beginning to close the gap between lab prototypes and commercial quantum computers.

Ultimately, quantum computing isn't just about the math—it's about the matter. The journey from theory to machine depends on the atoms, ions, photons, and superconductors we use to build these devices.

Understanding these architectures helps us see that quantum computing is not one technology—it's an **ecosystem of physics**, each with its own roadmap, risks, and revolutionary potential.

Chapter 9

Superconducting Qubits

In the global race to build scalable quantum computers, **superconducting qubits** have emerged as one of the most promising—and practical—technologies. Used by major players like IBM, Google, and Rigetti, this architecture is at the forefront of quantum hardware development.

At their core, superconducting qubits are electrical circuits built from **superconducting materials**, which exhibit zero electrical resistance when cooled to near absolute zero. These circuits are engineered to behave like artificial atoms, with discrete energy levels that can represent quantum information.

How Superconducting Qubits Work

The magic of superconducting qubits starts with the **Josephson junction**—a thin insulating barrier between two superconducting materials. Unlike classical inductors or capacitors, Josephson junctions introduce **nonlinearity** into the circuit, enabling the creation of well-separated quantum energy levels.

This nonlinearity allows engineers to isolate just two energy levels —typically the ground state |0⟩ and first excited state |1⟩—to act as a qubit, even though the system has many possible energy levels like a harmonic oscillator.

The most common design is the **transmon qubit**, a variant that reduces sensitivity to charge noise while maintaining good control over the qubit's frequency and coherence. Transmons are typically built on a silicon or sapphire substrate using standard lithographic processes, making them compatible with existing semiconductor fabrication techniques.

Once cooled inside a **dilution refrigerator** to temperatures around 10–20 millikelvin, these superconducting circuits behave quantum mechanically. At such temperatures, thermal energy is so low that the qubit remains in its ground state unless deliberately manipulated.

Controlling and Measuring Qubits

To perform quantum operations, superconducting qubits are controlled by **microwave pulses** sent through coaxial cables into the cryostat. These pulses manipulate the quantum state by rotating it around the Bloch sphere, effectively applying quantum gates like X, Y, Z, and Hadamard.

Two-qubit gates, such as **CNOT** or **iSWAP**, are implemented by temporarily coupling qubits together through bus resonators or tunable couplers. This allows for controlled interactions and entanglement.

Readout is typically performed by coupling each qubit to a **readout resonator**—a circuit whose resonance frequency shifts slightly depending on the qubit's state. By probing this resonator with a microwave tone and measuring the reflected signal, the state of the qubit can be inferred without directly collapsing its full wave function until the end of the computation.

Advantages of Superconducting Qubits

One of the biggest strengths of superconducting qubits is **speed**. Single-qubit gate times are measured in **tens of nanoseconds**, while two-qubit gates take just over 100 nanoseconds. These fast gates allow more operations to be completed within the limited coherence time.

Superconducting qubits also benefit from **mature fabrication infrastructure**. Many of the processes and tools developed for silicon-based microelectronics can be adapted for quantum chip manufacturing, allowing for faster iteration and integration.

The technology is **highly programmable** and well-supported by software ecosystems like IBM's Qiskit and Google's Cirq. With access to cloud-based superconducting quantum processors, developers can write and run quantum circuits on real hardware today.

Most importantly, superconducting qubits have demonstrated **quantum supremacy**—a milestone claimed by Google in 2019 using its 53-qubit "Sycamore" processor. Though limited to a contrived task, the experiment showcased that superconducting qubit systems could outperform the world's best classical supercomputers on specific quantum problems.

Limitations and Challenges

Despite their promise, superconducting qubits are far from perfect. One major limitation is **coherence time**—the amount of time a qubit can maintain its quantum state. In practice, coherence times range from 50 to 200 microseconds, meaning computations must be completed quickly before decoherence sets in.

Additionally, superconducting qubits are vulnerable to **crosstalk**, **noise**, and **material defects**. Even minor inconsistencies in fabrication or electromagnetic interference can lead to variability in qubit performance.

Another challenge is **scalability**. Each qubit requires its own control lines, readout circuitry, and isolation. As systems scale into the hundreds or thousands of qubits, managing the complexity of the wiring, thermal load, and calibration becomes a major engineering hurdle.

This has prompted research into **modular designs**, **3D integration**, and **cryogenic electronics** to reduce footprint and improve scalability. Companies are also exploring **qubit connectivity maps** and **quantum multiplexing** to make better use of hardware without adding more control lines.

Error Correction and the Path to Fault Tolerance

Superconducting qubits, like all physical qubit implementations, are error-prone. Quantum gates introduce small inaccuracies, and decoherence adds randomness to outcomes. To build useful quantum computers, **quantum error correction (QEC)** is essential.

The most promising approach is the **surface code**, a topological error-correcting code that spreads a logical qubit over a large grid of physical qubits. Surface code requires thousands of physical qubits for every error-corrected logical qubit, setting the bar high for hardware requirements.

Companies like IBM are developing **logical qubits** using their superconducting platforms and have published detailed roadmaps toward **fault-tolerant architectures**—predicting large-scale quantum systems within the next decade.

Industry Progress and Future Outlook

IBM, Google, and others have made significant strides in scaling up superconducting quantum systems:

- IBM's Quantum Eagle processor features **127 qubits** and is part of its roadmap to deliver a **1,000+ qubit machine** (Condor).
- Google's Sycamore system continues to refine control and reduce errors while exploring new layout and connectivity models.
- Startups like Rigetti are exploring modular "quantum tiles" that can be networked together to build larger machines.

Research into improving coherence, reducing crosstalk, and enhancing gate fidelity is ongoing. Advances in materials science, microwave engineering, and cryogenic packaging are essential to push performance beyond today's limits.

The future of superconducting qubits lies in **error-corrected quantum computing**. As hardware improves and error rates decline, these systems could begin tackling real-world problems in chemistry, machine learning, and optimization.

Promise and Precision

Superconducting qubits represent a pragmatic and fast-evolving path to quantum computing. They blend theoretical power with hands-on engineering, offering the clearest path today to medium- and large-scale quantum systems.

Their success reflects not only the elegance of quantum physics, but also the **precision of modern fabrication**, the **discipline of microwave engineering**, and the **vision of scalable architectures**.

While the road to fully fault-tolerant quantum computing remains steep, superconducting qubits have already taken us farther than many thought possible. As researchers refine this technology, the dream of practical quantum advantage inches closer to reality.

Chapter 10

Trapped Ions and Photonic Qubits

While superconducting qubits have led the early wave of commercial quantum computing, they are not the only contenders. Two alternative approaches—**trapped ions** and **photonic qubits**—offer radically different paths, each with unique advantages in fidelity, stability, and architecture.

These systems represent not just technical alternatives, but **philosophical contrasts** in how we build quantum computers: one relying on the **pristine isolation of matter**, and the other on the **speed and resilience of light**.

Trapped Ion Qubits: Precision from the Atomic Scale

Trapped ion quantum computers use **individual charged atoms**, typically ytterbium (Yb^+), barium (Ba^+), or calcium (Ca^+), to represent qubits. The quantum state is encoded in the **electronic energy levels** of these ions, which are manipulated using laser light.

To isolate and control the ions, they are suspended in **vacuum chambers** using electromagnetic fields generated by **Paul traps** or **Penning traps**. In this ultra-clean, ultra-stable environment, each ion hovers in place, forming a linear or two-dimensional array.

What makes trapped ions particularly powerful is their **exceptional coherence time**. Because they are identical particles of nature, each ion behaves exactly the same as the next. Coherence times in these systems can reach **seconds or even minutes**, vastly outperforming many other platforms.

Quantum gates in trapped ion systems are implemented with laser pulses that target specific ions. **Single-qubit gates** use focused beams to excite energy level transitions, while **two-qubit gates** typically involve coupling ions via shared vibrational modes, using techniques like the **Mølmer – Sørensen gate**.

Another major advantage is **connectivity**. In small ion chains, each qubit can interact with any other—a fully connected topology that's ideal for quantum algorithms requiring high qubit interdependence.

However, this architecture is not without its challenges. Scaling up trapped ion systems is difficult:

- **Vibrational modes** become harder to isolate as more ions are added.
- **Laser control** becomes more complex and error-prone with hundreds of optical paths.
- **Gate speeds** are relatively slow—microseconds for single-qubit gates and tens to hundreds of microseconds for two-qubit gates.

To overcome these limitations, companies like IonQ, Quantinuum, and Alpine Quantum Technologies are developing modular and shuttled architectures where ions can be moved or networked between zones for distributed computing.

Despite these hurdles, trapped ion systems remain a gold standard for **gate fidelity** and **repeatability**. They are widely used in quantum simulation, error correction research, and early commercial use cases where **precision is paramount**.

Photonic Qubits: Computing at the Speed of Light

While trapped ions are anchored in space, **photonic quantum computers** are built around **light**. Photonic qubits use properties of photons—such as **polarization**, **path**, **time-bin**, or **frequency**—to encode quantum information.

Photons are natural carriers of quantum states. They **don't interact with their environment** easily, which makes them highly resistant to decoherence. They can also travel long distances through optical fibers, making them ideal for **quantum communication** and **networked computing**.

Photonic qubits are manipulated through **beam splitters**, **phase shifters**, **interferometers**, and **nonlinear optical materials**. Measurements are done using single-photon detectors, which register the presence or absence of a photon in a given mode.

One of the leading techniques is **measurement-based quantum computing**, where quantum gates are performed not by direct interaction, but by **measuring entangled resource states** (known as **cluster states**) and conditioning subsequent operations on the outcomes.

This approach is **highly parallel** and well-suited for **integrated photonic chips**, where light paths are etched into silicon substrates and controlled with precision optics. Companies like **Xanadu**, **PsiQuantum**, and **Light matter** are pioneering scalable photonic quantum platforms using this technique.

The benefits of photonic systems include:

- **Room-temperature operation**
- **Inherent scalability** via chip integration
- **Low error from environmental noise**
- **Natural compatibility with fiber-optic infrastructure**

However, photonic quantum computing has its own steep challenges:

- **Generating indistinguishable single photons on demand** is technically difficult.
- **Photon-photon interactions** are weak, requiring probabilistic gates or specialized materials.
- **Losses** in optical elements and detectors can severely impact fidelity.

To address these, researchers are developing **quantum dot emitters**, **squeezed light sources**, and **heralded gate protocols**. Success in these areas could unlock scalable architectures with hundreds or thousands of photonic qubits, interconnected in modular arrays.

Photonic systems are particularly well-suited for **quantum networking**, **secure communications**, and **Boson sampling**—a specialized problem where quantum light can outperform classical simulations, offering a benchmark for near-term quantum advantage.

Comparing the Two Approaches

Feature	Trapped Ions	Photonic Qubits
Physical Qubit	Charged atom (ion)	Single photon
Physical Qubit	Lasers for gates	Linear optics + measurement
Environment	Ultra-high vacuum + cryogenics	Room temperature
Coherence Time	Seconds	Long (if no loss)
Gate Speed	Slow	Fast (but probabilistic)

Scalability	Moderate (shuttling/ modular)	High (chip-based)
Best For	Precision logic, simulation	Communication, networking

Both platforms challenge conventional engineering and computing models. Trapped ions offer **precision and fidelity** unmatched by most other systems, while photonic qubits promise **speed, scalability, and integration** with global quantum networks.

It's not yet clear which system will win in general-purpose quantum computing. In fact, the future may belong to **hybrid architectures**, where trapped ions perform precise logic and photonic qubits handle long-distance communication and entanglement distribution.

Diverse Paths to Quantum Power

Trapped ions and photonic qubits show us that there is no single blueprint for a quantum computer. The rules of the quantum world allow for **multiple physical embodiments** of information, each with strengths suited to different challenges.

What they share is ambition: to push beyond the limits of classical computing by harnessing the laws of nature at their most fundamental level.

Whether it's the stillness of atomic ions or the speed of light itself, these platforms represent humanity's boldest experiments in information processing. And they remind us that the quantum revolution will be just as diverse and dynamic as the technologies it inspires.

Chapter 11

Topological Qubits and Anyons

Topological qubits are an innovative and theoretically robust approach to quantum computing, offering a fundamentally different method for protecting quantum information. Unlike other types of qubits, which are vulnerable to local noise and errors, topological qubits derive their resilience from the global, topological properties of the system they inhabit. This robustness arises from encoding quantum information non-locally, which makes the qubit inherently immune to many forms of decoherence that typically plague quantum systems.

At the heart of topological quantum computing lies the concept of **anyons**, exotic quasi-particles that can exist in two-dimensional systems. Unlike familiar particles such as bosons and fermions, anyons obey fractional statistics, meaning that when two anyons are exchanged, the quantum state of the system changes in a way that depends on the order of the exchange. This non-Abelian braiding property is central to how topological qubits function and is the basis for performing quantum operations.

Anyons are predicted to occur in certain condensed matter systems, especially in the fractional quantum Hall effect and in topological superconductors. In these environments, particles are not elementary in the traditional sense but are emergent phenomena—collective excitations of many electrons that behave as if they were particles. These emergent particles can possess non-trivial statistics and topological characteristics.

Topological quantum computing is performed by **braiding anyons** —that is, moving them around each other in a two-dimensional plane. Each braid corresponds to a specific quantum gate, and because the result of the computation depends only on the topological class of the braid, the process is intrinsically error-tolerant. This is what sets topological qubits apart: their resistance to small local disturbances.

The most well-known theoretical model for topological qubits is the **Majorana zero mode**, which emerges at the ends of topological superconducting wires. These modes are their own antiparticles and can form non-Abelian anyons. When paired and manipulated appropriately, they can encode a qubit in a way that spreads the quantum information across space, making it less susceptible to local interference.

Implementing topological qubits in practice has proven immensely challenging, despite promising theoretical frameworks. The hunt for Majorana zero modes is ongoing, with several experiments reporting signatures that could indicate their presence, although definitive proof remains elusive. Materials like indium an timonide nanowires coupled with superconductors are among the leading candidates for realizing these modes.

The potential of topological qubits lies in their **fault-tolerant quantum computation**. In conventional quantum systems, error correction is necessary to maintain coherent computation, often requiring many physical qubits for each logical qubit. Topological qubits could, in principle, dramatically reduce the overhead needed for error correction due to their built-in protection mechanisms.

Topological quantum computers could enable more stable and scalable systems, but they currently lag behind other approaches such as superconducting qubits and trapped ions in terms of experimental realization and gate fidelity. However, they hold a key advantage in long-term scalability and reliability, assuming the underlying theoretical principles can be translated into practical devices.

Companies like Microsoft have invested heavily in topological quantum computing research, aiming to harness Majorana fermions for building more robust quantum hardware. Their strategy diverges from other leading quantum hardware developers, betting on the eventual success of topologically protected qubits over brute-force error correction.

From a physics perspective, anyons represent a fascinating frontier in our understanding of quantum statistics. Their existence challenges the traditional division of particles into bosons and fermions, revealing that nature allows for far richer behaviors in reduced dimensions. The mathematical framework that describes anyons draws from **topology, group theory, and quantum field theory**, making this field a vibrant intersection of theoretical physics and practical engineering.

Beyond computation, the study of anyons and topological phases of matter has broader implications for **condensed matter physics**. It has led to a deeper understanding of quantum states of matter, especially those that cannot be described using conventional symmetry-breaking theories. Topological phases are characterized by global properties, such as winding numbers or Chern numbers, that remain unchanged under continuous deformations.

The road ahead for topological qubits involves overcoming significant experimental barriers. Precise control of material properties, ultra clean fabrication techniques, and cryogenic systems are all necessary for stabilizing and detecting Majorana modes or other anyonic excitations. Moreover, new methods of manipulating and reading out the state of these qubits must be refined to make topological quantum computing viable.

Topological qubits and anyons represent one of the most elegant and theoretically grounded approaches to quantum computing. While practical implementation remains in its infancy, the promise of intrinsic fault tolerance and robust information encoding makes this field one of the most exciting areas in quantum research. Continued advances in material science, nano-fabrication, and quantum control will be critical in determining whether topological qubits can move from theory to reality in the coming years.

Chapter 12

Quantum Control and Calibration

Quantum control and calibration are foundational elements in the development and operation of quantum technologies. Whether in quantum computing, sensing, or communication, precise manipulation and stabilization of quantum systems are essential to ensure accuracy, reliability, and performance. Without robust control mechanisms and fine-tuned calibration protocols, quantum devices would quickly succumb to noise, decoherence, and operational errors.

At the core of quantum control is the ability to direct the evolution of quantum states using external fields, such as laser pulses, microwave signals, or magnetic fields. These controls must be carefully crafted to implement specific quantum gates, entangle qubits, or prepare quantum systems in desired states. Because quantum systems are highly sensitive, even small errors in timing, amplitude, or frequency can lead to significant deviations from expected behavior.

Quantum calibration, in turn, refers to the process of fine-tuning the parameters of quantum operations to achieve optimal performance. This involves characterizing system components, identifying error sources, and iteratively adjusting control signals to minimize those errors. Effective calibration enables high-fidelity quantum gates, reduces crosstalk between qubits, and ensures consistency across experimental runs.

The challenge of quantum control lies in the quantum system's susceptibility to environmental disturbances and the no-cloning theorem, which prevents the duplication of quantum states for error checking. As a result, quantum control strategies must operate in a regime where measurement is either indirect or destructive, requiring clever feedback mechanisms and adaptive protocols that make use of partial information.

Several control methods have been developed across different quantum platforms. In superconducting qubits, microwave pulses are shaped using optimal control techniques such as GRAPE (Gradient Ascent Pulse Engineering) or DRAG (Derivative Removal by Adiabatic Gate) to minimize leakage and non-ideal transitions. In trapped ions, laser pulse sequences are optimized to reduce motional excitation and off-resonant coupling.

Machine learning and artificial intelligence are increasingly being applied to quantum calibration. Reinforcement learning agents can explore the control parameter space more efficiently than manual tuning, finding optimal sequences that maximize gate fidelity or suppress unwanted dynamics. These automated calibration tools are crucial for scaling quantum systems to larger numbers of qubits, where manual methods become infeasible.

Control fidelity is typically quantified using benchmarks such as randomized benchmarking or quantum process tomography. These tools help assess the accuracy of quantum gates and calibrations by statistically sampling the output of many random operations. From this, the average fidelity can be determined, offering insight into how closely the implemented operations match their theoretical ideals.

Drift and noise are persistent problems in quantum hardware, making continuous calibration a necessity. Many quantum systems require recalibration multiple times per day, as temperature fluctuations, electromagnetic interference, or mechanical vibrations can subtly alter the system dynamics. Therefore, real-time feedback loops and autonomous recalibration are key features of modern quantum platforms.

Calibration also involves characterizing qubit parameters like coherence times (T1 and T2), anharmonicity, and frequency detuning. These parameters must be monitored regularly to ensure consistent behavior across time and devices. In systems like

Trapped Ions

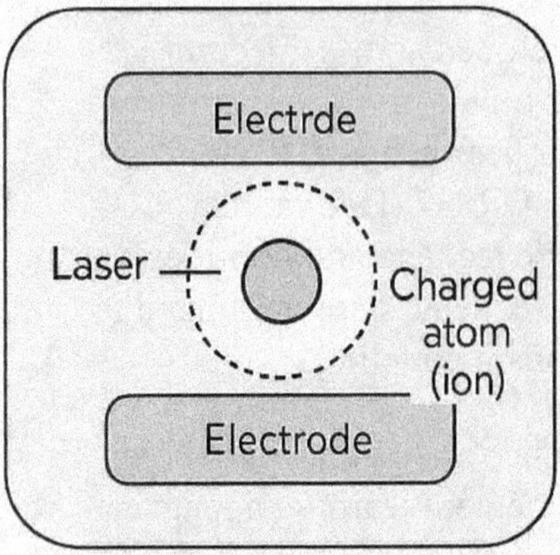

Photonic Qubits

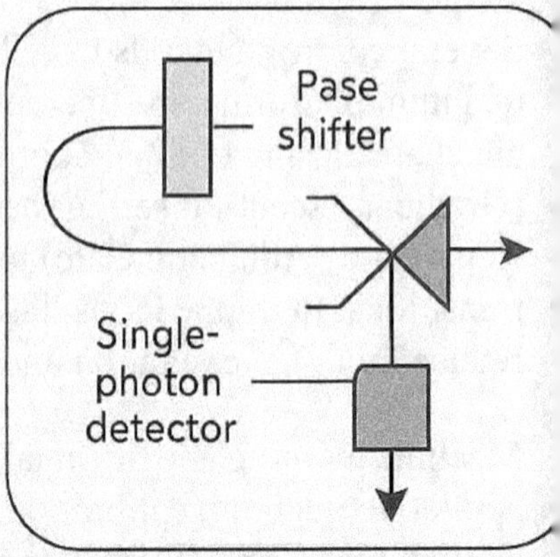

	Trapped Ions	Photonic Qubits
Physical Qubit	Charged atom (ion)	Single photon
Operation	Lasers for gates	Linear optics + measurement
Environment	Uitra-high vacuum + cryogenics	Room temperature
Coherence Time	Seconds	Long (if no loss)
Gate Speed	Slow	Fast (but probabilistic)
Scalability	Moderate (shuttling/ modular)	High (chip-based)
Best For	Precision logic, quantum simulation	Communication, networkin

---

The physical qubit, identifies thraopica lizes laser for gates, while Photonic Qubits use linear optics combined with measurement.

---

superconducting circuits, flux-tuning and frequency sweet spots are used to optimize performance while avoiding parasitic interactions.

For multiqubit systems, crosstalk and correlated noise present additional control challenges. Isolating and controlling individual qubits becomes harder as systems scale. Advanced pulse shaping, frequency multiplexing, and dynamic decoupling techniques are employed to mitigate these effects, enabling the implementation of two-qubit and multi-qubit gates with higher fidelity.

In quantum annealers and analog quantum simulators, control focuses more on slowly evolving system parameters to guide the system toward desired ground states. Calibration in these systems involves tuning annealing schedules, bias fields, and coupling strengths to align with theoretical models or desired computational outputs.

The interface between quantum control and hardware engineering is crucial. High-quality control requires precision electronics, stable timing systems, and low-noise environments. Microwave generators, arbitrary waveform generators, and field-programmable gate arrays (FPGAs) play essential roles in delivering precise control signals with minimal latency.

Quantum control is not only about precision but also about speed. Faster gates reduce exposure to decoherence, but must be balanced against the risk of inducing errors from spectral leakage or unwanted transitions. This trade-off drives ongoing research into pulse optimization, control theory, and quantum hardware design.

Ultimately, control and calibration are the keys to unlocking the full potential of quantum technology. They determine the quality of quantum computation, the accuracy of quantum sensors, and the reliability of quantum communication protocols. As quantum systems scale and become more complex, the need for automated, adaptive, and robust control strategies becomes even more pressing.

The development of standardized calibration protocols and control software is already underway. Open-source tools and software frameworks—like Qiskit, Cirq, and Py-Quil—offer modules for calibration and pulse-level control, enabling researchers and engineers to implement complex control logic and tailor it to their specific hardware.

In the future, quantum control will likely become increasingly abstracted from the user, with high-level programming environments automatically translating logical instructions into optimal control sequences. This evolution parallels the history of classical computing, where early users had to interact with machines at the hardware level, but eventually moved to high-level languages and compilers.

As we move closer to fault-tolerant quantum computing, the role of control and calibration will remain central. Even in error-corrected systems, gates must be performed with high enough fidelity to stay below the fault-tolerance threshold. This necessitates continued advances in control algorithms, hardware engineering, and calibration methodologies.

In summary, quantum control and calibration form the invisible scaffolding of quantum technology. Without them, even the most advanced quantum hardware would fail to deliver meaningful results. Through a combination of physical insight, algorithmic innovation, and engineering precision, quantum control enables us to steer fragile quantum systems toward practical, powerful applications.

Section 3: Quantum Error Correction

Chapter 13

The Need for Error Correction

Quantum computing holds the promise of solving certain problems exponentially faster than classical computers. However, this promise is tempered by the fragile nature of quantum information. Unlike classical bits, which are robust and can be easily checked and corrected, quantum bits—or qubits—are susceptible to a wide range of errors. These include bit flips, phase flips, decoherence, and operational imperfections. As a result, **quantum error correction (QEC)** is not a luxury, but a **fundamental necessity** for building scalable and reliable quantum computers.

One of the main reasons quantum systems are error-prone is their sensitivity to the environment. Qubits interact with surrounding electromagnetic fields, nearby particles, or even thermal noise, causing their quantum state to degrade. This phenomenon, known as decoherence, can destroy the superposition and entanglement needed for quantum computations. Moreover, because quantum mechanics forbids directly copying unknown quantum states (the no-cloning theorem), standard classical error correction techniques cannot be directly applied.

Unlike classical bits that are either 0 or 1, qubits can exist in a superposition of both. This makes detecting and correcting errors more complex. A key insight in quantum error correction is that while you can't clone quantum information, you can **spread it across multiple physical qubits** in such a way that the information can be recovered even if some qubits are corrupted. This forms the basis of quantum error correction codes, which protect a single logical qubit using multiple physical qubits.

The most well-known quantum error correction code is the **Shor code**, which protects against both bit and phase errors using a 9-qubit encoding. Since then, more efficient codes have been developed, such as the **Steane code**, **surface code**, and **color code**, each offering trade-offs in terms of overhead, fault tolerance, and implementation complexity. Among these, the surface code is considered the most promising for near-term fault-tolerant quantum computers due to its tolerance of high error rates and compatibility with two-dimensional qubit architectures.

Quantum error correction is not just about detecting and fixing errors—it's about **building fault-tolerant systems** where computations can proceed reliably even as errors occur. Fault-tolerant quantum computing refers to architectures and protocols that ensure computations remain accurate despite the presence of noise, as long as the error rates are below a certain threshold. This is known as the **fault-tolerance threshold**, and it is one of the central concepts in quantum error correction theory.

Achieving fault-tolerance requires implementing a full **quantum error correction cycle**, which includes encoding, syndrome measurement, error identification, and recovery, all while avoiding the introduction of new errors. Syndrome measurements are performed in such a way that they reveal information about the error—not the quantum state itself—preserving the superposition and entanglement needed for computation.

One of the significant challenges in quantum error correction is the **resource overhead**. To protect even a single logical qubit, one may need dozens or even hundreds of physical qubits. For example, surface code implementations often require more than 1,000 physical qubits to encode a single high-fidelity logical qubit. This makes large-scale quantum error correction currently impractical for the small, noisy devices available today—often referred to as **Noisy Intermediate-Scale Quantum (NISQ)** devices.

Nonetheless, the development of error correction techniques is progressing rapidly. Experimental demonstrations of small-scale error correction—such as stabilizing a qubit using three- or five-qubit codes—have shown promising results. Researchers have successfully implemented basic correction cycles in systems like superconducting qubits, trapped ions, and photonic qubits.

Error correction also plays a role beyond computing. In **quantum communication**, for instance, quantum repeaters and entanglement purification protocols rely on similar ideas to overcome loss and noise in transmitting quantum information across long distances. Similarly, in **quantum sensing**, error correction can help extend coherence times and improve measurement precision.

Ultimately, the need for error correction is what distinguishes experimental prototypes from truly scalable quantum computers. Without error correction, quantum systems can perform small computations or simulations but will be limited in scope and reliability. With robust error correction, quantum computers can execute long, complex algorithms and unlock applications in cryptography, chemistry, optimization, and beyond.

In summary, quantum error correction is essential to ensure the **viability, scalability, and trustworthiness** of quantum technologies. It transforms inherently noisy quantum systems into powerful, fault-tolerant computing platforms. Although the technical challenges are steep, advances in theory, hardware, and control techniques are steadily paving the way toward practical error-corrected quantum computing.

Chapter 14

Surface Codes and Stabilizer Formalism

Surface codes are among the most promising and practically implementable forms of quantum error-correcting codes. Their structure and functionality are deeply rooted in the **stabilizer formalism**, a powerful mathematical framework used to describe and manipulate quantum error correction codes in an efficient and scalable way. Together, surface codes and stabilizer techniques represent one of the leading approaches to achieving **fault-tolerant quantum computing**.

The **stabilizer formalism** was developed to generalize and simplify the description of quantum error-correcting codes. Instead of explicitly tracking the state of every qubit, the stabilizer formalism describes quantum codes through a set of **commuting operators** (called stabilizers) that define the code space. These stabilizers are tensor products of Pauli operators (I, X, Y, Z), and the valid quantum states in the code are the simultaneous +1 eigenstates of all the stabilizer operators.

In this framework, errors can be detected by measuring stabilizers: if a stabilizer yields a –1 eigenvalue, it indicates that an error has occurred on one or more of the associated qubits. Crucially, stabilizer measurements reveal information about the **error syndrome** without collapsing the encoded quantum information. This non-destructive property is essential for continuous error detection and correction.

The **surface code** is a two-dimensional lattice-based stabilizer code that arranges qubits in a grid and defines stabilizers locally. There are two main types of stabilizers in the surface code: **plaquette (Z-type)** operators, which check for phase flip (Z) errors, and **star (X-type)** operators, which check for bit flip (X) errors. Each stabilizer involves four neighboring qubits arranged around a plaquette or star, enabling detection of errors with minimal locality requirements.

One of the primary advantages of the surface code is its **high fault-tolerance threshold**—often quoted between 0.5% and 1% error rates per gate, depending on implementation. This makes it one of the most practical and scalable codes for near-term and future quantum processors. Additionally, because it only requires nearest-neighbor interactions on a two-dimensional grid, it aligns well with the physical layout constraints of platforms like superconducting qubits and quantum dots.

The logical qubits in a surface code are not individual physical qubits but rather **non-local constructs** defined by extended paths (strings) across the lattice. Logical operators correspond to chains of X or Z operations that connect boundaries or wind around defects (holes) in the lattice. These extended operators are what provide the surface code's intrinsic robustness to local errors: to corrupt a logical qubit, an error chain must span the entire code, which becomes exponentially unlikely with increasing code distance.

Code distance in the surface code is defined as the minimum number of physical errors required to cause a logical error. A higher code distance improves error resilience but requires more physical qubits. For example, a distance-3 surface code can detect and correct any single-qubit error, while a distance-5 code can handle up to two.

Implementing surface codes requires **syndrome extraction**, the process of repeatedly measuring stabilizers to detect errors without disturbing the encoded logical state. This is done using ancillary measurement qubits that interact with data qubits and are then measured to infer stabilizer values. These measurements must be repeated cyclically and efficiently decoded using classical algorithms to determine the most likely error and apply corrections.

Decoding the surface code is a classical problem that involves finding the most probable configuration of errors given the measured syndromes. Algorithms like **minimum weight perfect matching (MWPM)** are used to perform this task. Although decoding adds computational overhead, the locality and simplicity of the surface code's structure make it tractable for large systems.

The surface code also supports **fault-tolerant gate operations**. While single-qubit Clifford gates (like the Hadamard or phase gate) can often be implemented via lattice rotations or code deformation, non-Clifford gates such as the T gate require more elaborate procedures like **magic state distillation**. This necessity arises from the Eastin-Knill theorem, which prohibits the existence of a universal set of transversal gates for error-correcting codes.

Logical operations in the surface code can also be performed by **braiding defects** or **lattice surgery**. Braiding involves moving code defects around each other to implement logical gates, mimicking the conceptual idea of topological quantum computing. Lattice surgery, by contrast, joins and splits code patches to perform entangling operations between logical qubits, offering a modular and hardware-friendly approach.

Because of their locality, scalability, and error threshold, surface codes are the leading choice for many large-scale quantum computing efforts. Companies such as Google, IBM, and Quantinuum are actively working to implement and scale surface code-based architectures. Experimental milestones have included stabilizer measurement, logical qubit initialization, and simple forms of error correction.

Despite its strengths, surface code implementation is not without challenges. High-quality gate operations, fast and accurate measurements, and reliable ancilla qubits are all necessary for successful error correction. Additionally, magic state distillation, needed for universal computation, imposes significant overheads, prompting research into more efficient protocols and alternative code constructions.

In summary, the combination of **surface codes and stabilizer formalism** offers a mathematically elegant and experimentally feasible foundation for fault-tolerant quantum computing. Stabilizer theory simplifies the understanding and implementation of quantum error correction, while surface codes provide a concrete, scalable model that aligns with modern qubit architectures. As quantum hardware matures, the surface code will likely play a central role in bridging the gap between noisy devices and practical quantum applications.

Chapter 15

Fault-Tolerant Quantum Computing

Fault-tolerant quantum computing is the discipline and design principle that allows quantum computations to proceed reliably even in the presence of noise, decoherence, and gate errors. It is a cornerstone of making quantum computing scalable, practical, and useful for real-world applications. While today's quantum hardware is limited by relatively high error rates and short coherence times, fault-tolerant quantum computing aims to build systems that can operate indefinitely and correctly, provided errors remain below a certain threshold.

At the heart of fault tolerance is the recognition that physical qubits—those implemented in superconducting circuits, trapped ions, photonics, or other hardware platforms—are inherently noisy. Operations on these qubits are prone to errors from environmental interactions, imperfect control, and quantum measurement noise. Unlike classical computing, where error correction is relatively straightforward, quantum error correction is complicated by the no-cloning theorem and the fragile nature of quantum states.

To overcome these limitations, **fault-tolerant architectures encode logical qubits** into multiple physical qubits using quantum error-correcting codes. These logical qubits are designed so that small errors on individual physical qubits can be detected and corrected without disturbing the underlying quantum information. Surface codes, color codes, and concatenated codes are among the leading frameworks used for this purpose.

A system is considered **fault-tolerant** if it can detect and correct errors faster than they accumulate, enabling arbitrarily long computations. This requires not just error correction, but also fault-tolerant implementation of all operations: quantum gates, measurements, state preparation, and error syndromes must be executed in ways that do not propagate uncorrectable errors. For instance, a single error in a physical gate must not spread to multiple qubits in a way that would overwhelm the error-correcting code.

Fault tolerance also introduces the concept of a **threshold theorem**, which states that if the error rate per operation is below a certain threshold, arbitrarily reliable quantum computation is theoretically possible with sufficient overhead. This threshold varies depending on the specific code and error model, but for surface codes—a leading candidate for implementation—it's around 1%.

Achieving fault tolerance involves several layers:

1. **Encoding**: Logical qubits are encoded into multiple physical qubits using a quantum error-correcting code.

2. **Syndrome extraction**: Measurements of stabilizers or parity checks are used to identify potential errors.

3. **Decoding and correction**: Classical algorithms infer the likely error and apply corrections accordingly.

4. **Fault-tolerant gates**: Special protocols ensure that quantum gates applied to logical qubits do not introduce unmanageable errors.

Some gates, such as the Clifford group (e.g., CNOT, Hadamard, and Phase), can often be implemented transversally—by applying the same operation to each physical qubit in the code block independently. These gates are relatively easy to make fault-tolerant. However, **non-Clifford gates**, such as the T gate, are more difficult and typically require **magic state distillation**, which is resource-intensive but essential for achieving universal quantum computation.

One of the challenges of fault-tolerant quantum computing is the **overhead in physical resources**. A single logical qubit may require hundreds or thousands of physical qubits, depending on the code distance and target error rate. Despite this, continued improvements in gate fidelity, error correction algorithms, and qubit coherence times are reducing this overhead and moving fault-tolerant systems closer to reality.

In practice, fault-tolerant quantum computing also demands robust classical infrastructure. Fast and accurate decoding of syndromes, real-time control systems, and error-aware scheduling of operations all require tight integration between quantum and classical subsystems. These components must operate on time scales compatible with qubit coherence to ensure effective correction.

Researchers are actively exploring alternative architectures and techniques to improve fault tolerance. These include topological qubits (which aim to be inherently fault-tolerant), bosonic codes (that encode information in harmonic oscillators), and lattice surgery (a method to perform logical operations in surface codes with lower overhead).

Fault tolerance is more than just a technical requirement—it is a gateway to unlocking the full power of quantum computation. With fault-tolerant systems, it becomes possible to run deep quantum algorithms such as Shor's factoring, large-scale quantum simulations, and quantum machine learning models without the results being corrupted by noise. It also lays the groundwork for future quantum networks and distributed quantum computing.

In conclusion, **fault-tolerant quantum computing transforms error-prone quantum systems into reliable computational platforms**, bridging the gap between experimental physics and large-scale quantum applications. It is one of the most active and critical areas of quantum information science, drawing on deep insights from quantum theory, coding theory, computer architecture, and control systems. As hardware and theory continue to advance together, fault-tolerant quantum computing will be the foundation upon which practical, universal quantum computing is built.

Chapter 16

Decoding Algorithms and Thresholds

Decoding algorithms and thresholds are critical components in the framework of quantum error correction and fault-tolerant quantum computing. While quantum error-correcting codes protect information by distributing it across multiple physical qubits, decoding algorithms are tasked with interpreting the results of stabilizer measurements to identify and correct errors. The threshold, meanwhile, defines the maximum tolerable error rate per quantum operation below which quantum error correction becomes effective and scalable.

In classical error correction, decoding is relatively straightforward: parity checks detect where errors occurred, and corrections are applied directly. In quantum systems, however, things are more complex. Measurement cannot directly reveal the state of a quantum bit without collapsing it, so **syndrome measurements** are used instead. These indirect measurements provide clues about the presence and type of errors—bit flips (X), phase flips (Z), or both (Y)—without disturbing the quantum information encoded in the logical qubit.

A **decoding algorithm** takes the measured syndromes as input and attempts to infer the most likely set of errors that could have produced them. Its goal is to apply corrective operations that restore the logical qubit to its intended state. A key challenge is that many different error patterns can produce the same syndrome, especially as the number of qubits and possible errors increases. Thus, decoding becomes a probabilistic inference problem.

One of the most widely used decoding algorithms, especially for surface codes, is the **Minimum Weight Perfect Matching (MWPM)** algorithm. This algorithm treats error correction as a graph problem: the syndrome outcomes are represented as nodes, and the decoder finds the most likely connections (error paths) between them, assuming that the lowest-weight (i.e., fewest qubit) errors are most probable. MWPM is efficient and works well for uncorrelated, independent noise models, making it a practical choice in many quantum systems.

Another approach to decoding is **belief propagation**, which is well-known in classical coding theory. While it can be adapted to quantum codes, it performs less reliably for highly entangled or topological codes like the surface code. More recently, **machine learning-based decoders**, including neural networks and reinforcement learning agents, have been proposed. These techniques can potentially adapt to complex and correlated noise models, learning from real experimental data rather than assuming a simplified noise structure.

Thresholds play a central role in determining the viability of a quantum error-correcting code. The **fault-tolerance threshold** is the maximum error rate below which the error correction protocol can reduce the overall error probability as the code size increases. Above this threshold, increasing the number of physical qubits actually makes things worse; below it, the encoded logical qubits become more robust with scale.

Each code has a different threshold, and the threshold also depends on the noise model and the decoder used. For the surface code, one of the most studied and experimentally feasible codes, the threshold is typically around **1% per gate operation** under a simple depolarizing noise model and MWPM decoding. More advanced decoders and error models may change this value slightly.

Meeting or exceeding the threshold is a major milestone for any quantum computing platform. It means that the physical qubits are good enough to support reliable logical operations with reasonable overhead. Once this condition is satisfied, fault-tolerant architectures can, in theory, scale indefinitely, limited only by resources like the number of physical qubits and the efficiency of decoding.

In practice, decoding must be **fast and scalable**. In real-time quantum processors, stabilizer measurements are repeated frequently—potentially every microsecond—and decoding must keep up without introducing latency that would degrade system performance. This has driven the development of **hardware-accelerated decoders**, including those implemented on FPGAs and GPUs.

Furthermore, as quantum hardware becomes more complex and errors become correlated or context-dependent, **noise-aware decoding** becomes increasingly important. Decoders that assume uniform, independent noise may not perform optimally on actual devices, where crosstalk, leakage, or calibration drift can introduce complex error patterns. Adaptive and data-driven decoders are being actively researched to address this issue.

Decoding also ties into **syndrome extraction fidelity**. If the stabilizer measurements themselves are noisy, decoders must factor in the possibility of false syndromes. This has led to techniques such as repeated syndrome extraction, decoding across multiple rounds of measurement (space-time decoding), and incorporating measurement errors into the decoding algorithm itself.

In summary, decoding algorithms and thresholds are pivotal in determining whether quantum error correction can fulfill its promise. Efficient, accurate decoders make it possible to use imperfect hardware to protect fragile quantum information, while thresholds serve as benchmarks for the hardware quality needed to enable scalable, fault-tolerant quantum computing. Continued innovation in decoding strategies—classical, algorithmic, and machine learning-based—will be essential as quantum processors grow in size and complexity.

Chapter 17

Decoding Algorithms and Thresholds

Logical Qubits and Encoding Strategies

As quantum computers grow in power, so too does the challenge of keeping their calculations accurate. Qubits—the raw computing units of quantum systems—are inherently fragile. They suffer from **decoherence**, **environmental noise**, and **imperfect gates**. Even the slightest error can render a computation useless.

To perform meaningful, large-scale quantum computation, we must go beyond physical qubits. We must build **logical qubits**: error-resilient abstractions that behave like ideal, perfect qubits, even when made from imperfect parts.

Creating logical qubits is a central goal of **quantum error correction (QEC)**. Through clever encoding strategies, we can protect quantum information from errors—bit flips, phase flips, or both—without violating the no-cloning theorem or collapsing the quantum state.

This chapter explores what logical qubits are, how they're built, and why they're the foundation of **fault-tolerant quantum computing**.

What Is a Logical Qubit?

A **logical qubit** is a virtual, error-corrected qubit constructed from many physical qubits. While a physical qubit might decohere within 100 microseconds, a properly encoded logical qubit can maintain its state for much longer by detecting and correcting errors on the fly.

This process works by spreading quantum information across a set of entangled physical qubits in such a way that local errors can be identified and reversed—without directly measuring the logical state itself.

The number of physical qubits required to create a single logical qubit varies by code and error rate, but typical estimates range from **1,000 to 10,000 physical qubits per logical qubit** using today's noisy qubits.

Types of Quantum Errors

To understand encoding strategies, it helps to review the types of errors we aim to correct:

- **Bit-flip errors (X errors)**: A $|0\rangle$ becomes $|1\rangle$ or vice versa.
- **Phase-flip errors (Z errors)**: A $|+\rangle$ becomes $|-\rangle$—flipping the sign of the superposition.
- **Combined bit and phase flips (Y errors)**: A mix of both types simultaneously.

Quantum error correction must detect and correct all three types of errors while preserving coherence—a nontrivial task, given the restrictions of quantum measurement.

Encoding Strategies

1. Shor Code (9-Qubit Code)

The **Shor code** was one of the first quantum error-correcting codes. It encodes one logical qubit into nine physical qubits and corrects for any single-qubit error.

It works by first encoding the qubit redundantly across three groups to protect against bit-flips, then using entangled states to protect against phase-flips. Though foundational, it is not efficient for large-scale systems due to overhead.

2. Steane Code (7-Qubit Code)

The **Steane code** uses only seven physical qubits per logical qubit and can correct any single-qubit error. It is based on classical Hamming codes and uses stabilizer formalism to track and fix errors.

This code demonstrated how **quantum parity checks**—analogous to classical error detection—can be adapted to protect quantum information. It's more compact than Shor's and inspired many modern QEC codes.

3. Surface Codes

The **surface code** is the leading candidate for fault-tolerant quantum computing, widely studied and implemented by IBM, Google, and others.

It arranges physical qubits on a 2D lattice and uses **stabilizer measurements** to monitor error syndromes over time. Surface codes can detect and correct local errors with high efficiency and are well-suited to architectures with nearest-neighbor connectivity, like superconducting qubits.

Key benefits include:

- High error thresholds (up to ~1%)

- Local operations only (ideal for real hardware)
- Scalability via tiling of logical qubit blocks

However, they require many physical qubits—typically **1,000 or more per logical qubit**, depending on the desired logical error rate.

4. Color Codes

Color codes are a generalization of surface codes, using a different tiling pattern (triangular rather than square) and allowing for certain gates to be implemented more easily.

They support **transversal gates**, meaning logical gates can be implemented by applying the same gate across all physical qubits —useful for fault-tolerant logic.

Color codes are still under active research, especially in photonic and ion-trap implementations.

5. Concatenated Codes

Concatenation is a modular strategy where small QEC codes are nested inside larger ones. For example, each qubit in a 7-qubit Steane code could itself be encoded with a 3-qubit bit-flip code.

While powerful in theory, concatenated codes can be complex and require a large number of layers to reach low logical error rates.

Syndrome Measurement and Error Detection

Quantum error correction relies on **syndrome measurements**—indirect observations that detect errors without collapsing the quantum information. These are performed using **ancilla qubits** that interact with the encoded qubits, extract error information, and are then measured.

By interpreting the syndrome outcomes, the system can infer whether and where an error has occurred—and correct it using recovery operations.

This requires **real-time feedback**, precise calibration, and fast control electronics—challenges that are now being addressed by cutting-edge cryogenic and control hardware.

Fault-Tolerant Gates and Thresholds

A QEC code becomes **fault-tolerant** when it can not only detect and correct errors but also **prevent the spread of errors** during gate operations.

This requires careful circuit design to avoid uncorrectable multi-qubit faults. Some codes allow for **transversal gates**, where a logical operation is performed using only local, parallel gates—offering natural fault tolerance.

A key metric in these systems is the **error threshold**: the maximum error rate under which error correction is beneficial. For surface codes, this threshold is relatively high (~1%), making them feasible for today's imperfect hardware.

From Prototype to Practicality

Logical qubits are already being demonstrated on early hardware platforms:

- IBM is working on **small logical qubits** using surface code tiles.
- Google has tested **surface code cycles** for sustained error detection.

- Startups are exploring modular hardware optimized for encoding strategies.

The road to **practical logical qubits** will require:

- Higher coherence physical qubits
- Better gate fidelity
- Optimized layouts for syndrome extraction
- Low-latency control loops

When these elements converge, logical qubits will unlock the full potential of quantum computing—allowing circuits to run for hours, days, or longer without corruption.

Conclusion: Toward Fault-Tolerant Quantum Machines

Physical qubits are the raw material of quantum computers—but **logical qubits are the structure** that will support real-world applications. Through sophisticated encoding strategies, we can insulate quantum information from noise, enabling deeper circuits and more reliable results.

In classical computing, redundancy is taken for granted. In quantum computing, redundancy must be carefully engineered to preserve the delicate interplay of superposition, interference, and entanglement.

By learning how to encode, detect, and correct errors without destroying information, we aren't just building smarter machines—we're mastering quantum coherence itself.

The dream of a fault-tolerant quantum computer begins here, one logical qubit at a time.

Comparison of Quantum Error-Corre-Correcting Codes

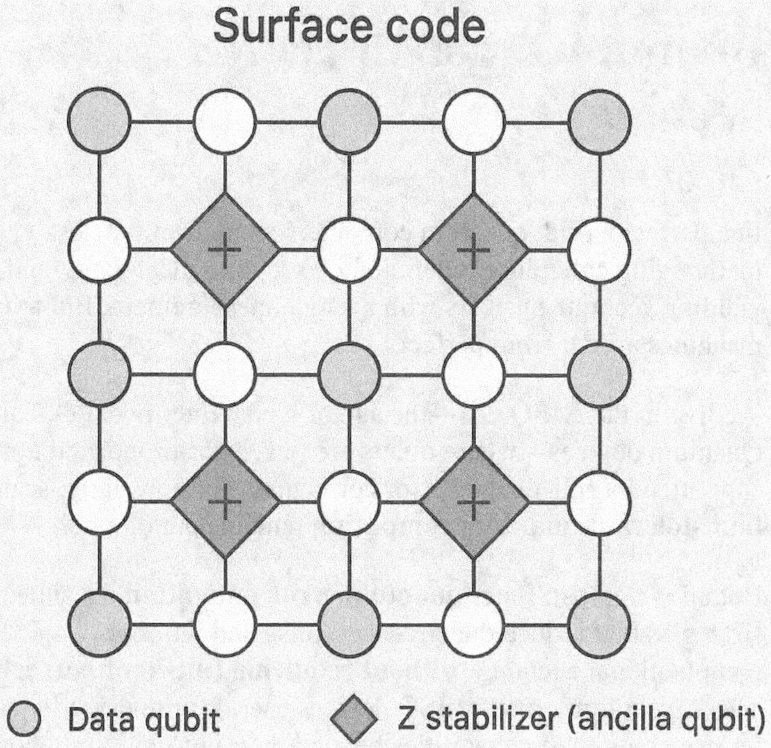

Chapter 18:

Quantum Error Mitigation Techniques

In a perfect world, quantum computers would run flawlessly, maintaining coherence indefinitely, executing ideal gates, and yielding accurate answers with a single measurement. But today's machines are far from perfect.

We live in the **NISQ era**—the age of Noisy Intermediate-Scale Quantum devices—where qubits are few, error-prone, and not yet supported by full-fledged error correction. For now, large-scale **fault-tolerant quantum computing** remains out of reach.

But all is not lost. Enter **quantum error mitigation**—a suite of strategies that reduce the effects of noise and improve computational accuracy **without requiring full error correction**. These techniques make NISQ devices useful for real-world tasks by stretching their capabilities beyond their physical limitations.

While mitigation doesn't prevent errors, it **compensates** for them. The goal is not to fix the machine, but to extract the best possible results from imperfect computations.

Why Error Mitigation?

Quantum error correction (QEC) promises long-term resilience, but it comes with enormous overhead. Most codes require hundreds or thousands of physical qubits per logical qubit—well beyond current hardware limits.

Mitigation techniques, in contrast, work **within today's constraints**. They assume a small number of qubits and relatively high noise, and operate by:

- Characterizing the noise,
- Modifying circuits or measurements,
- Post-processing results to reduce bias.

They're especially valuable for variational quantum algorithms like **VQE** and **QAOA**, which run short circuits repeatedly and rely on statistical output.

1. Zero-Noise Extrapolation (ZNE)

One of the most widely used techniques, **Zero-Noise Extrapolation** involves artificially increasing the noise in a circuit and using multiple runs to estimate what the result would be with **no noise**.

Here's how it works:

- The circuit is executed at normal noise (baseline).
- Then, noise is deliberately amplified by stretching gate durations or repeating gates.
- The results are fit to a curve (e.g., linear or exponential), and extrapolated back to estimate the zero-noise outcome.

ZNE works well for shallow circuits and observables with smooth noise scaling. It does not require changes to the hardware—only smart repetition and post-processing.

2. Probabilistic Error Cancellation

This technique is the closest analog to a **virtual error correction**. The idea is to learn a **noise model** of the device, then reverse it mathematically by applying an inverse transformation during post-processing.

Steps include:

- Characterize the noise on gates via **quantum process tomography** or **Pauli twirling**.
- Express the noisy gate as a mixture of ideal gates and known errors.
- Sample from this mixture to statistically **cancel** the noise's effect.

While highly effective in principle, error cancellation is **resource-intensive**. It can increase sampling requirements exponentially and depends on highly accurate noise models.

3. Clifford Data Regression (CDR)

CDR uses **easy-to-simulate Clifford circuits** (those that can be classically computed) as references. It runs these circuits on the real quantum device and compares the noisy results to the known correct answers.

This comparison builds a regression model that can predict the correct answer for **harder, non-Clifford circuits** that are not classically simula table.

CDR is an example of **machine learning applied to quantum noise**, using known data to calibrate and correct unknown computations. It's promising but still experimental and data-hungry.

4. Measurement Error Mitigation

A surprisingly large portion of quantum error arises not during computation, but during **readout**. Measurement errors occur when a |0⟩ is reported as |1⟩, or vice versa.

To mitigate this:

- A **calibration matrix** is built by measuring known basis states multiple times.
- This matrix models the noise in the measurement process.
- Results are **corrected (inverted)** by applying the inverse of the matrix to the observed outputs.

This technique is relatively low-cost and widely used, especially in NISQ platforms like IBM Quantum and Rigetti.

5. Subspace Expansion

Used in quantum chemistry and VQE, **subspace expansion** corrects errors by expanding the variational ansatz into a larger space of basis states.

This space includes not just the computed state, but nearby states that can be accessed via simple operations (e.g., applying Pauli operators).

By solving a generalized eigenvalue problem in this subspace, more accurate energies or expectation values can be extracted—effectively smoothing out local noise effects.

6. Dynamical Decoupling

Although not mitigation in the post-processing sense, **dynamical decoupling** helps preserve qubit coherence during idle periods.

This technique applies a sequence of carefully timed pulses to a qubit to **refocus** its state and cancel out low-frequency noise, similar to a spin echo in NMR.

It's especially useful in hardware where gate time is fast, but idle qubits experience drift and noise.

7. Learning-Based Mitigation

Recent advances are using **machine learning models**—like neural networks or kernel methods—to learn patterns in noisy output distributions and map them back to noise-free estimates.

While promising, this approach requires large datasets and careful generalization, as overfitting can introduce new biases.

As quantum data volumes grow, this hybrid **quantum-classical feedback loop** will become a more robust tool for intelligent error control.

Challenges and Trade-offs

While mitigation is powerful, it comes with costs:

- **Increased sampling**: Many mitigation techniques require more circuit repetitions.
- **Limited scalability**: Most work best with small circuits or shallow depths.
- **Assumption-heavy**: Techniques like ZNE or error cancellation assume specific noise models that may not hold in all hardware.
- **No long-term solution**: Mitigation cannot scale indefinitely—it delays but doesn't eliminate the need for true error correction.

Nonetheless, in the NISQ era, these techniques are the **only practical defense** against quantum error. They enable experiments, benchmarking, and early applications that would otherwise be meaningless under raw noise.

Conclusion: Making the Most of NISQ

Quantum error mitigation represents **the art of doing more with less**. In a world where perfect qubits are a distant dream, mitigation allows researchers to extract signal from noise, and precision from imperfection.

By understanding the types of errors in a system—and tailoring strategies to reduce their impact—we make quantum devices useful today, even as we build toward the fault-tolerant machines of tomorrow.

For anyone working in quantum algorithm development, chemistry simulations, or variational optimization, mitigation is not an optional add-on—it is **a core tool of the trade**.

In the chapters ahead, we'll explore how these techniques are applied in real-world quantum applications, and how mitigation integrates with classical post-processing and hardware calibration to push the frontier of what NISQ computers can do.

Section 4: Quantum Algorithms

Chapter 19

Introduction to Quantum Algorithms

A quantum computer, like any computer, is only as powerful as the algorithms it runs. But unlike classical algorithms, which operate on bits and deterministic logic, **quantum algorithms** operate on qubits, probability amplitudes, superpositions, and entanglement.

This shift in paradigm leads to algorithms that are not only faster for certain tasks, but fundamentally different in their structure, logic, and resource usage.

Quantum algorithms are designed to **exploit the unique properties of quantum mechanics**—such as superposition, interference, and entanglement—to solve problems more efficiently than their classical counterparts. Some of these algorithms offer polynomial speedups. Others offer exponential breakthroughs.

But before diving into individual examples, it's essential to understand what makes a quantum algorithm **quantum** in the first place.

What Is a Quantum Algorithm?

At a high level, a quantum algorithm is a procedure that runs on a **quantum circuit**. It takes as input a set of qubits in some initial state, evolves them through a series of **quantum gates**, and finally extracts information through **measurement**.

What makes quantum algorithms powerful is the way they structure computation as a **probability distribution over many possible outcomes**, rather than a single deterministic path. By designing circuits that **amplify the right outcomes and suppress the wrong ones**, quantum algorithms arrive at solutions efficiently.

A successful quantum algorithm must:

- Use **superposition** to explore many paths simultaneously,
- Use **entanglement** to link variables in ways classical computers can't,
- Use **interference** to guide the computation toward correct results,
- Use **measurement** to extract those results meaningfully.

Why Do We Need Quantum Algorithms?

Quantum algorithms are not faster at *everything*. In fact, many tasks can be done just as well—or better—on classical machines.

But for certain classes of problems, quantum algorithms offer **dramatic improvements**:

- **Integer factorization**, classically intractable at large scales, becomes feasible with Shor's algorithm.
- **Unstructured search** is sped up quadratically by Grover's algorithm.
- **Quantum simulation** of molecules and materials, impossible to simulate accurately on classical computers, is naturally suited to quantum hardware.
- **Optimization**, **machine learning**, and **linear algebra** are also seeing quantum-inspired advances.

Key Features of Quantum Algorithms

1. Superposition-Based Parallelism

Quantum algorithms explore multiple states at once via superposition. A single n-qubit register can represent 2^n states simultaneously. This "massive parallelism" isn't directly observable, but it enables fast transformations of state space.

2. Interference

Rather than storing every result, quantum algorithms **engineer interference patterns** to amplify the right answers and cancel out wrong ones—an elegant method of computation by design, not by brute force.

3. Probabilistic Output

Quantum algorithms are **inherently probabilistic**. Results may need to be sampled multiple times to identify the most likely (correct) answer. This adds complexity but also flexibility, particularly in noisy or heuristic problems.

4. Reversibility and Unitarity

Every operation in a quantum algorithm must be **reversible** and described by a unitary matrix. This constraint leads to interesting designs that are quite unlike traditional programs, and often require auxiliary "ancilla" qubits.

Categories of Quantum Algorithms

Quantum algorithms can be grouped into broad categories based on the types of problems they address:

1. Algorithms for Number Theory and Cryptography

- **Shor's Algorithm**: Efficiently factors large integers and breaks RSA encryption.

- **Quantum Discrete Logarithm**: Solves another hard classical problem related to cryptographic security.

2. Search and Optimization

- **Grover's Algorithm**: Searches an unstructured database of N entries in $\sqrt{N}$ time.
- **Quantum Approximate Optimization Algorithm (QAOA)**: A variational hybrid algorithm used for solving combinatorial problems like Max-Cut.

3. Quantum Simulation

- **Trotter-Suzuki and Hamiltonian Simulation**: Simulates quantum systems over time.
- **Variational Quantum Eigensolver (VQE)**: Computes ground-state energies of molecules—critical in quantum chemistry.

4. Machine Learning and Linear Algebra

- **Quantum Principal Component Analysis (qPCA)**
- **Quantum Support Vector Machines**
- **HHL Algorithm**: Solves systems of linear equations exponentially faster under specific conditions.

5. Sampling and Probabilistic Models

- **Boson Sampling**: Demonstrates quantum advantage with photonic systems, though not yet useful in practice.
- **Quantum Walks**: Generalize classical random walks, often with speedups in graph-related problems.

Hybrid Quantum-Classical Algorithms

Because today's quantum machines are limited in size and fidelity, many algorithms are designed to work in **hybrid mode**—splitting the workload between classical processors and quantum devices.

Examples include:

- **VQE**: A classical optimizer adjusts quantum circuit parameters to minimize an observable.
- **QAOA**: A classical loop guides quantum sampling toward optimal solutions.

These hybrid algorithms are **modular**, noise-resilient, and especially well-suited for NISQ devices.

Designing a Quantum Algorithm

Developing a new quantum algorithm is challenging. It requires:

- Deep understanding of quantum mechanics and linear algebra,
- Familiarity with unitary transformations and entanglement patterns,
- Creativity in designing circuits that guide amplitude evolution,
- Simulations and optimization tools to test designs before implementation.

Fortunately, tools like **Qiskit**, **Cirq**, **Pennylane**, and **Braket** allow researchers and developers to write, test, and execute quantum algorithms on real hardware or simulators.

Limitations and Caveats

Quantum algorithms are not **universally faster**. They often rely on assumptions—such as structured input, low noise, or specific oracle access. For many real-world problems, no known quantum advantage exists (yet).

Moreover, algorithms must be robust to **hardware limitations**: decoherence, gate errors, and limited qubit connectivity. This places practical constraints on what can be computed today.

But in areas where quantum advantage is possible, quantum algorithms offer not just speed—but **new ways of thinking about computation**.

The Quantum Algorithmic Mindset

Quantum algorithms are more than fast versions of classical ones. They represent a **new logic**, shaped by the laws of quantum physics. Designing them requires a shift in mindset—from deterministic step-by-step procedures to **wave-like, probabilistic, amplitude-guided evolution**.

As hardware matures, the demand for quantum algorithms will explode—across industries from finance to logistics to drug design.

This chapter has laid the foundation. In the next chapters, we'll explore some of the most famous and impactful quantum algorithms—starting with **Shor's Algorithm for factoring** and **Grover's Algorithm for search**—and see how they actually work in circuits.

The future of quantum computing depends not just on qubits, but on **what we do with them**. Algorithms will be the key to unlocking that future.

Chapter 20

Shor's Algorithm and Integer Factoring

Few algorithms in quantum computing have generated as much interest—or concern—as **Shor's algorithm**. Proposed by mathematician **Peter Shor** in 1994, it introduced a quantum method for **factoring large integers exponentially faster** than any known classical algorithm.

At its core, Shor's algorithm is a **threat to classical encryption** systems, particularly **RSA**, which secures the majority of online communication and commerce today. RSA's security rests on the difficulty of factoring the product of two large prime numbers—a task that would take classical computers millions of years for keys of practical size.

Shor's algorithm doesn't just speed this up. It **redefines the problem**, exploiting quantum mechanics to reduce the runtime from **exponential** to **polynomial** time. This theoretical breakthrough marked the beginning of quantum computing's rise from abstract theory to national security priority.

The Integer Factoring Problem

Let's define the task:

Given a large composite number **N** (e.g., 2048-bit RSA key), find two non-trivial integers **p** and **q** such that **N = p × q**.

Classically, the best known general-purpose algorithm—**the General Number Field Sieve (GNFS)**—has **sub-exponential** complexity. But it still becomes infeasible for numbers with more than a few hundred digits.

Shor's quantum algorithm can factor **N** in **polynomial time**, roughly **$O((\log N)^3)$**, making RSA and other public-key systems fundamentally vulnerable in a post-quantum world.

The Big Idea: Reduce Factoring to Period Finding

Rather than factoring directly, Shor's algorithm transforms the problem into one of **period finding**, which is where quantum computing excels.

Here's the clever approach:

1. **Choose a random integer** *a*, such that $1 < a < N$ and **gcd(a, N) = 1** (i.e., a is co-prime to N).
2. Define the function **$f(x) = a^x \bmod N$**. This function is **periodic** with some unknown period **r**.
3. Find the **period r**, such that **f(x + r) = f(x)** for all x.
4. If **r is even** and **a^(r/2) ≠ -1 mod N**, then:
   - One can compute **gcd(a^(r/2) ± 1, N)**, which yields a non-trivial factor of N.

The bottleneck is **finding r**, which is classically hard—but can be done **efficiently using a quantum Fourier transform** (QFT) on a superposition of inputs.

How Shor's Algorithm Works (High-Level Steps)

1. **Initialize Registers**:

- Two quantum registers are used. The first holds a superposition of inputs **x**, and the second stores the function output **f(x) = a^x mod N**.

2. **Create Superposition**:

   - Apply Hadamard gates to the first register to create a superposition of all integers from 0 to Q–1, where Q is a power of 2 greater than N^2.

3. **Modular Exponentiation**:

   - Compute **a^x mod N** in the second register for all **x** in superposition. This entangles the input and output, preparing for interference.

4. **Apply Quantum Fourier Transform**:

   - Apply the QFT to the first register. This transforms the uniform superposition into one that highlights the **periodicity** of f(x) in its frequency domain.

5. **Measure and Classical Post-Processing**:

   - Measure the first register to obtain a value that encodes a multiple of **1/r**.

   - Use **continued fractions** classically to estimate **r**, then test if it yields a valid factor.

6. **Repeat if Necessary**:

   - The process may need to be repeated a few times to get a suitable period **r**.

Quantum vs Classical: Why It's Revolutionary

The key quantum advantage comes from using **interference patterns** to extract hidden structure in the modular exponential function—a task that is intractable classically due to the exponential number of function evaluations required.

Using the quantum Fourier transform, Shor's algorithm turns this massive search into a **constructive measurement**, allowing a small quantum circuit to recover the period with high probability.

While the algorithm contains classical steps (e.g., computing gcd, continued fractions), the heavy lifting—period finding—is uniquely quantum.

Impact on Cryptography

Shor's algorithm shatters the assumption that factoring is hard—an assumption that underpins:

- **RSA**
- **DSA**
- **Diffie-Hellman**
- **Elliptic Curve Cryptography (ECC)**

With a large enough fault-tolerant quantum computer, these protocols would be **breakable in polynomial time**, rendering much of today's internet security obsolete.

This has triggered a global race toward **post-quantum cryptography (PQC)**, with NIST and other agencies developing classical algorithms resilient to quantum attacks.

Current Limitations

Shor's algorithm is proven, but **not yet practical** for large numbers because:

- It requires **modular exponentiation circuits**, which are deep and complex.
- It demands **thousands of logical qubits**, far beyond what current noisy devices can sustain.
- Quantum error correction must be implemented to run the algorithm at useful scale.

The largest number factored by Shor's algorithm on a real quantum device remains small (e.g., **21, 15, 35**)—mostly proofs of concept.

But as hardware improves and **logical qubits** become viable, Shor's algorithm will transition from theoretical threat to active capability.

Conclusion: A Glimpse into Quantum Power

Shor's algorithm isn't just a powerful algorithm—it's a symbol. It represents the potential of quantum computing to **redefine computational boundaries**, break old assumptions, and rewrite the rules of digital security.

More than 25 years since its invention, Shor's algorithm still drives quantum hardware development, cryptographic reform, and research into fault-tolerant architectures.

It shows us that quantum computing is not just about simulating molecules or speeding up search. It's about **fundamentally changing what's computable**.

In the next chapter, we'll explore **Grover's Algorithm**, which, while offering a smaller speedup, applies to a much broader class of problems—including search, optimization, and quantum-enhanced AI.

The following visual is a summary for Shor's Algorithm—including a high-level circuit diagram and a sidebar on cryptographic implications.

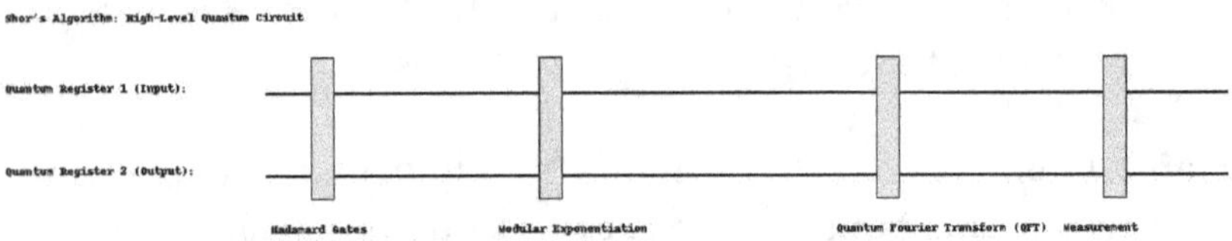

Cryptographic Implications of Shor's Algorithm:

- Breaks RSA and ECC encryption in polynomial time.
- Threatens public-key infrastructure globally.
- Forces migration to Post-Quantum Cryptography (PQC).
- Drives national security and NIST standards efforts.
- Currently limited by hardware constraints but inevitable long-term.

Chapter 21

Grover's Algorithm for Unstructured Search

While Shor's algorithm targets a specific mathematical problem—integer factoring—**Grover's algorithm** applies to a much broader and more common class of problems: **search**.

Originally developed by **Lov Grover** in 1996, this algorithm addresses the question:

"Given an unstructured database of N items, how can we find a marked item (a solution) as efficiently as possible?"

Classically, this would require **O(N)** time—checking each item one by one until the correct one is found. But Grover's quantum algorithm can do it in just **O(√N)** steps. This may sound modest compared to Shor's exponential speedup, but it represents the **best possible quantum speedup for unstructured search**, and applies to a huge range of practical problems.

From optimization and scheduling to pattern recognition and machine learning, Grover's framework has become one of the cornerstones of quantum algorithm design.

The Problem: Searching a Black Box

Suppose we have a function **f(x)** that maps inputs {0, 1, ..., N−1} to either 0 or 1, where **f(x) = 1** if x is a solution and 0 otherwise.

We are promised that **at least one solution exists**. Our task is to find an x such that **f(x) = 1**, but we know nothing about the structure of f. We can only query it—this is an **oracle model**.

In classical computing, we'd need **O(N)** queries in the worst case. Grover's algorithm reduces this to roughly **O(√N)**queries, achieving a **quadratic speedup**.

Core Concepts of Grover's Algorithm

Grover's algorithm revolves around two powerful ideas:

1. **Amplitude Amplification**:
   Instead of flipping a bit when we find a solution, we **rotate the quantum amplitude** toward the desired answer, so that measuring the state is more likely to yield the correct result.

2. **Oracle Query and Diffusion Operator**:
   The algorithm uses two main components: an **oracle** that marks the solution by flipping its phase, and a **diffusion operator** (also called the Grover iterate) that inverts all amplitudes around their average—effectively amplifying the marked state.

Step-by-Step: How Grover's Algorithm Works

1. Initialize the Qubits

We start with n qubits in state $|0\rangle$ and apply Hadamard gates to each one, creating a uniform superposition over all **2^n**possible inputs:

$\|\psi\rangle = N\ 1\ x=0 \Sigma N-1\ \|x\rangle$

This is our **initial state**, where each possible input has an equal amplitude.

2. Oracle Function (Phase Inversion)

Next, we use a **quantum oracle**, a black-box operation that flips the phase of the solution state:

$U_f|x\rangle=(-1)^{f(x)}|x\rangle$

This means: for the correct solution x, $f(x) = 1$, and its amplitude is **negated**, while all other amplitudes remain unchanged.

3. Diffusion Operator (Inversion About the Mean)

After the oracle marks the solution, we apply the **diffusion operator**, which reflects the state vector about the average amplitude. This increases the probability amplitude of the solution and decreases others.

This step ensures that over several iterations, the probability of measuring the correct solution increases significantly.

Mathematically, it's implemented as:

$D=2|\psi\rangle\langle\psi|-I$

This is a form of **constructive interference**—the essence of quantum computation.

4. Repeat (Grover Iterations)

We repeat the oracle + diffusion steps about **$\sqrt{N}$** times. Each iteration rotates the quantum state closer to the correct answer. Too few or too many rotations will decrease accuracy, so knowing the number of solutions or estimating it is crucial.

5. Measurement

Finally, we measure the state. Because of the amplitude amplification, the correct solution has the **highest probability** of being observed.

An Example: Search Over 4 Elements

Say N = 4. Classically, we need up to 4 queries to find the solution.

With Grover's algorithm:

- Start with 2 qubits → 4 possible states: $|00\rangle, |01\rangle, |10\rangle, |11\rangle$.
- Apply Hadamard's → uniform superposition.
- Apply oracle: flip sign of target state.
- Apply diffusion → amplify target state.
- After one iteration, measurement yields the correct answer with high probability.

Complexity and Optimality

- **Query complexity**: Grover's algorithm uses $O(\sqrt{N})$ oracle calls.
- **Gate complexity**: Depends on the complexity of implementing the oracle and the diffusion operator.
- **Optimality**: It's provably the best quantum algorithm for unstructured search—no algorithm can do better in the black-box model.

Variants and Applications

Grover's framework can be adapted for many applications:

- **Multiple solutions**: Adjust iteration count accordingly.
- **Quantum counting**: Estimate the number of solutions.

- **Amplitude estimation**: Used in quantum machine learning and Monte Carlo integration.

Practical applications include:

- Cryptanalysis (e.g., brute-force breaking of symmetric encryption).
- Constraint satisfaction problems.
- Search in large datasets or configuration spaces.
- Variants of machine learning (e.g., k-means initialization).

Grover vs Shor: Different Domains, Equal Impact

While Shor's algorithm is more famous due to its implications for cryptography, Grover's algorithm is **more broadly applicable**. Its quadratic speedup may not sound revolutionary, but in massive search spaces (e.g., cryptographic keys or combinatorial problems), it can make a real difference.

Importantly, Grover's algorithm works with **NISQ-era devices** for small problems and serves as a blueprint for hybrid quantum-classical search and optimization strategies.

Conclusion: Amplifying Answers in the Quantum Way

Grover's algorithm is a beautiful demonstration of how **quantum mechanics reshapes computation**. It doesn't brute-force through possibilities—it **evolves amplitudes** to make the right answer rise to the top.

Its elegance lies in its simplicity and generality: a few steps, a

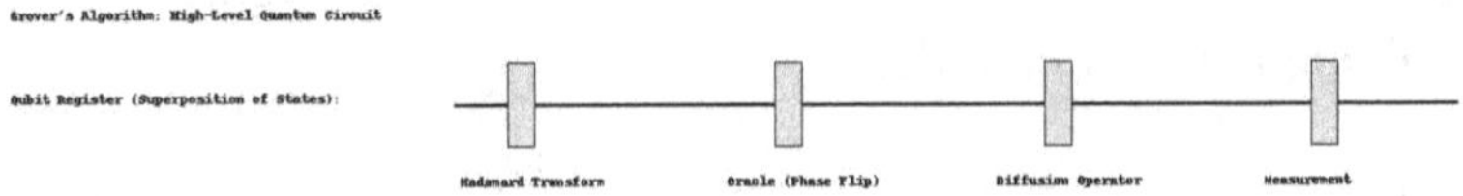

Grover's Algorithm vs Classical Search:

- Classical Search: O(N) time complexity
- Grover's Quantum Search: O(sqrt(N)) complexity
- Example: Searching 1 item in 1 million takes ~1000 steps classically, ~100 with Grover
- Ideal for cryptanalysis, unstructured data, and large search spaces
- Works on NISQ-era devices with small search spaces

clever use of reflection and interference, and a provable improvement over classical limits.

As quantum computers scale, Grover's algorithm will become a go-to tool for quantum-accelerated search, optimization, and decision-making—quietly but powerfully amplifying the solutions that matter.

The following visual is a summary for Grover's Algorithm—including the high-level circuit diagram and a sidebar comparing it with classical search

Chapter 22

Quantum Fourier Transform

If quantum algorithms had a magic wand, it would be the **Quantum Fourier Transform (QFT)**.

Just as the classical **Fourier Transform** decomposes signals into frequencies, the QFT decomposes quantum states into **amplitude frequencies**—allowing quantum computers to extract periodicity, interference patterns, and hidden structure.

This transform is not just a mathematical curiosity; it is central to **Shor's algorithm**, **quantum phase estimation**, and many applications in **signal processing**, **chemistry**, and **cryptography**. Understanding the QFT is essential to grasping the power and architecture of quantum algorithms.

What Is the Quantum Fourier Transform?

The **Quantum Fourier Transform** is the quantum analogue of the classical Discrete Fourier Transform (DFT). It maps a quantum state in the computational basis to a new basis of phase states.

Given an *n*-qubit state:

∥x⟩=k=0∑2n−1 ak ∥k⟩

The QFT maps it to:
QFT(∥x⟩)=y=0∑2n−1 a~y ∥y⟩

Where:
a~y =2n 1 x=0∑2n−1 ax ⋅e2πixy/2n

This transformation is unitary, reversible, and computable by a **quantum circuit of polynomial depth**.

Why the QFT Matters

In quantum computing, we often need to:

- Detect **periodicity** (as in Shor's algorithm),
- Estimate **phases** (as in quantum phase estimation),
- Perform **signal analysis** on quantum states.

The QFT enables all of this with **exponential efficiency**. Classically, a DFT on 2^n inputs takes $O(2^n \log 2^n)$ time. The QFT can be performed with **$O(n^2)$** quantum gates—exponentially faster.

Structure of the QFT Circuit

The QFT circuit is elegant and recursive. For an *n*-qubit system, it consists of:

1. **Hadamard Gate on the First Qubit**
2. **Controlled Phase Rotations** on subsequent qubits:
    - R_k gates apply phase shifts of $2\pi/2^k$.
3. **Repetition** for each qubit, moving down the line.
4. **Swap Gate Layer** to reverse the qubit order.

Example for 3 Qubits:

- Apply **Hadamard** to qubit 0.
- Apply $\mathbf{R_2}$ to qubit 1 controlled by qubit 0.

- Apply $\mathbf{R_3}$ to qubit 2 controlled by qubit 0.
- Move to qubit 1: apply Hadamard, R_2 to qubit 2.
- Finally, Hadamard on qubit 2.
- **Swap** qubits 0 and 2.

This process spreads amplitude information across the entire register using **phase encoding**.

Controlled Phase Rotation Gates

These are key to the QFT and take the form:

Rk =[10 0e2πi/2k]

They introduce precise phase differences between basis states, which allows the quantum state to **interfere constructively or destructively** in a controlled way.

Inverting the QFT

Because the QFT is unitary, it has a well-defined inverse: **QFT†**.

QFT†=QFT−1

In many algorithms—such as phase estimation—we apply the QFT, then measure, then apply QFT† to backtrack or un-compute ancilla registers. This symmetry makes the QFT a reusable and modular tool.

Applications of the QFT

1. Shor's Algorithm

The QFT is used to extract the period **r** of a function via frequency estimation. This is what enables integer factoring in polynomial time.

2. Quantum Phase Estimation (QPE)

QPE uses the QFT to convert phase information into a binary register—a critical step in quantum chemistry, Hamiltonian simulation, and more.

3. Hidden Subgroup Problems

Many algorithms in the **BQP** class use QFT over various groups to find hidden structures efficiently.

4. Quantum Signal Processing

Emerging fields like **quantum sensing** and **quantum machine learning** may leverage QFTs to compress or transform data encoded in amplitude or phase.

Efficiency and Practical Considerations

- The full QFT uses $O(n^2)$ gates, but an **approximate QFT**—omitting tiny-angle phase gates—can reduce complexity without much error.
- Controlled-phase gates are **difficult on noisy hardware**, but manageable on fault-tolerant architectures.
- QFT circuits are relatively shallow, which makes them suitable for NISQ devices when truncated.

Feature	Classical FFT	Quantum Fourier Transform

Input Size	N (real/complex values)	2^n (quantum amplitudes)
Time Complexity	O(N log N)	$O(n^2)$
Reversibility	Optional	Always unitary
Implementation	Software/hardware	Quantum circuits
Applications	Signal processing	Period finding, QPE, cryptography

The Heart of Quantum Periodicity

The QFT is more than a mathematical transformation—it is **a lens into the hidden structure of quantum states**.

Where classical algorithms loop through inputs, the QFT **extracts global information** in a single sweep, using interference to reveal patterns that would be invisible to classical eyes.

As quantum algorithms evolve, the QFT remains one of the most powerful primitives, enabling breakthroughs in factoring, chemistry, optimization, and beyond.

It exemplifies what makes quantum computing unique—not just faster, but **deeper**, able to see **into the wave** rather than across it.

The following labeled diagram is a **3-qubit Quantum Fourier Transform (QFT) circuit**, along with a sidebar summarizing its role in various quantum algorithms

Quantum Fourier Transform (QFT) - 3 Qubit Circuit

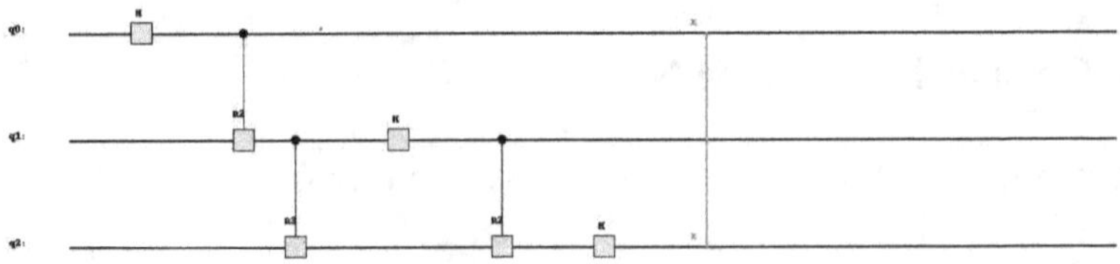

QFT in Quantum Algorithms:

- Shor's Algorithm: Period finding for factoring integers.
- Quantum Phase Estimation: Extract phase information from eigenstates.
- Hidden Subgroup Problems: Detect hidden symmetries in functions.
- Quantum Signal Processing: Encode and manipulate phase-encoded signals.
- Hamiltonian Simulation: Useful in simulating time evolution of quantum systems.

Chapter 23

Amplitude Amplification Techniques

In classical computing, we rely on **probabilities** to reason about uncertainty. In quantum computing, we rely on **amplitudes**—complex numbers that determine the probability of observing a quantum state.

Quantum amplitudes aren't just passive descriptors—they can be **engineered and amplified** to favor desired outcomes. This powerful concept lies at the heart of **amplitude amplification**, a technique that significantly boosts the likelihood of measuring correct or "good" solutions in quantum algorithms.

Where **Grover's algorithm** is a specific instance tailored to unstructured search, amplitude amplification is the **generalized framework** that applies to a broader class of quantum algorithms and probabilistic procedures.

From Probability to Amplitude

Consider a quantum algorithm **A** that prepares a state:

$$A|0\rangle = p\,|\psi_{good}\rangle + 1 - p\,|\psi_{bad}\rangle$$

Where:

$|\psi_{good}\rangle$ contains the desired outcome(s),
$|\psi_{bad}\rangle$ represents incorrect or unhelpful outcomes,
p is the probability of getting a good result upon measurement.

Amplitude amplification aims to **increase the amplitude of** $|\psi_{good}\rangle$ so that we get the right result with high probability, using a process similar to Grover's iterative rotations in Hilbert space.

The Core Procedure

Let **A** be a unitary quantum algorithm that generates a superposition over possible solutions. Let **Q** be the Grover operator:

Q=–AS0 A−1Sf

Where:

- Sf : Phase flip of the "good" states (oracle marking good states),
- S0 : Phase flip of the |0⟩ state,
- A−1: Uncomputes the initial preparation,
- Q: Amplifies the amplitude of good states via interference.

By applying **Q** repeatedly (~O(1/√p) times), we rotate the initial state toward |ψgood ⟩, increasing the probability of success.

Key Properties

- **Quadratic Speedup**: If a classical algorithm has success probability *p*, and we repeat it O(1/p) times to get a good outcome, amplitude amplification improves this to O(1/√p) iterations.

- **Black-box compatible**: The method works even if we don't know the structure of *A*—we just need it to be a quantum circuit.

- **Reversibility required**: The initial state preparation must be unitary and reversible to define the reflection operators properly.

Relation to Grover's Algorithm

Grover's search is a special case of amplitude amplification where:

- *A* is the Hadamard transform creating uniform superposition,
- *S_f* is the oracle flipping the phase of the solution,
- $p = 1/N$ for a single solution in an unstructured space.

Thus, amplitude amplification **generalizes Grover's method** to any algorithm that probabilistically prepares good solutions.

Applications Beyond Search

Amplitude amplification enhances many probabilistic quantum procedures, including:

1. Quantum Monte Carlo

By applying amplitude amplification to sampling procedures, one can estimate integrals and expected values more efficiently than classical Monte Carlo methods.

2. Decision Problems

If a quantum procedure can detect a property (e.g., whether a solution exists), amplitude amplification can boost its reliability exponentially.

3. Machine Learning

In quantum-enhanced classification, amplitude amplification can improve the probability of selecting the right class or label after a variational training procedure.

4. Quantum Chemistry

Used within **variational algorithms** (like VQE), where finding the ground state often involves preparing a likely-good solution and amplifying it toward convergence.

Quantum Circuit View

1. A typical amplitude amplification circuit contains:
2. State preparation (unitary *A*)
3. Oracle/marking operation Sf
4. Un-computation A-1
5. Reflection about the initial state S0
6. Repetition of this process

Each repetition increases the amplitude of the "good" state like rotating a vector closer to its target in a 2D subspace of the Hilbert space.

\Trade-offs and Considerations

- **Over-rotation**: Too many iterations cause the amplitude to **overshoot** and reduce accuracy. This makes knowing or estimating *p* crucial.

- **Noise sensitivity**: Amplitude amplification involves coherent, iterative steps—susceptible to decoherence in NISQ devices.

- **Oracle cost**: If the oracle or A

  A–1 is expensive, the amplification becomes less efficient in practice.

Still, when *p* is small, the **quadratic gain** in success probability is invaluable.

Hybrid Use with Classical Post-Processing

Many quantum-classical hybrid algorithms (e.g., QAOA or VQE) can embed amplitude amplification in their inner loops, particularly in settings where repeated quantum measurements yield noisy, probabilistic answers.

By focusing amplitude on likely solutions, we reduce the **classical sampling burden** and improve the efficiency of the overall pipeline.

Amplifying Success in Quantum Computation

Amplitude amplification shows that quantum advantage isn't always about exponential speedups. Sometimes, it's about being **smart with uncertainty**—and using the structure of quantum amplitudes to push us closer to the answers we need.

As a general technique, it equips quantum developers with a powerful method to **enhance accuracy, reliability, and performance**, regardless of the specific problem domain.

In the quantum world, getting the right answer isn't just about more computing—it's about the **right interference**. And amplitude amplification is how we tune that interference in our favor.

Amplitude Amplification Circuit (Generalized Grover's Algorithm)

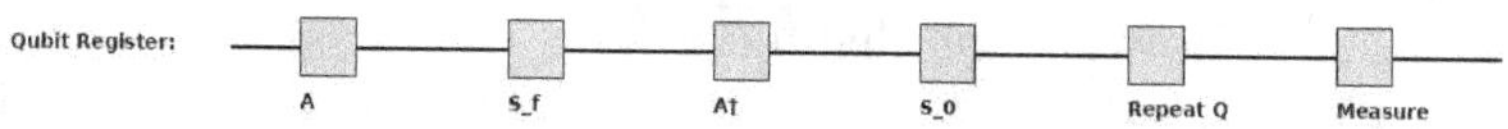

Amplitude Amplification vs Classical Repetition vs Grover's:

- Classical Repetition: Requires O(1/p) iterations for success probability p.
- Amplitude Amplification: Achieves same with O(1/sqrt(p)) iterations.
- Grover's Algorithm: Special case for uniform superposition and unique solution.
- Amplification works with any initial state prep algorithm (A).
- Gains quadratic speedup while remaining broadly applicable.

Chapter 24

Quantum Walks and Simulation Algorithms

When we think of algorithms, we often imagine instructions moving forward, step by step, like a person walking down a path. In classical computing, this idea takes form in **random walks**—stochastic processes that model diffusion, search, or even finance. In quantum computing, this idea is elevated and reimagined as a **quantum walk**, where the evolution is not random but coherent, unitary, and governed by the laws of quantum mechanics.

Quantum walks have emerged as a rich and versatile tool in the quantum algorithm toolbox, offering new ways to solve graph problems, simulate quantum systems, and understand quantum dynamics.

In parallel, **simulation algorithms**—especially those targeting quantum systems themselves—have become a primary motivation for building quantum computers in the first place. Richard Feynman once envisioned quantum systems as fundamentally beyond classical simulation. With quantum walks and related methods, we are now closing that loop.

What is a Quantum Walk?

A **quantum walk** is the quantum counterpart of a classical random walk, where a "walker" moves over a graph or lattice structure. However, instead of random transitions, a quantum walk uses **unitary operations** to evolve a quantum state.

There are two main types:

- **Discrete-time quantum walks (DTQW):** Require both a "coin" and position space to direct motion.
- **Continuous-time quantum walks (CTQW):** Evolve directly on the graph's adjacency matrix using the Schrödinger equation.

The quantum walk explores paths in **superposition**, and the resulting interference patterns can significantly outperform classical walks in terms of mixing time, hitting time, and search efficiency.

Discrete-Time Quantum Walk (DTQW)

A DTQW has two registers:

- **Coin register**: Determines the direction of movement (like flipping a quantum coin).
- **Position register**: Tracks the walker's location on the graph or line.

The walk proceeds in steps:

1. **Coin flip** (Hadamard or other unitary on the coin qubit),
2. **Conditional shift** (move left or right depending on coin state),
3. Repeat.

This causes probability amplitude to **interfere**, spreading out faster than classical probability distributions and enabling faster search or mixing.

Continuous-Time Quantum Walk (CTQW)

In CTQW, no coin is required. The evolution of the system is governed by:

$idtd \, |\psi(t)\rangle = H|\psi(t)\rangle$

Where **H** is the **graph Laplacian or adjacency matrix** acting as a Hamiltonian.

These walks are widely used in **quantum simulation**, especially in modeling particle dynamics or quantum transport on physical networks.

Applications of Quantum Walks

1. Search Algorithms on Graphs

Quantum walks can find marked vertices in a graph exponentially faster than classical counterparts under specific conditions. Examples include:

- **Element distinctness**: O(N^2/3) vs. O(N) classically.
- **Triangle finding** in graphs.

2. NAND Tree Evaluation

Quantum walks can evaluate Boolean formulas embedded in tree structures more efficiently, showing an early separation between quantum and classical query complexities.

3. Spatial Search

Locating marked locations on a 2D or 3D grid can be accelerated with quantum walk dynamics, especially with well-designed boundary conditions.

4. Quantum Simulation

CTQWs model particle dynamics, quantum chaos, or even chemical reactions, where amplitude flow between quantum states follows the physical structure of the problem.

5. Quantum Machine Learning

Quantum walks form the basis for algorithms like **quantum kernel estimation** or **graph-based learning**, where similarity and connectivity are naturally encoded in quantum dynamics.

Quantum Simulation Algorithms

Beyond quantum walks, **simulation algorithms** aim to evolve quantum systems under given Hamiltonians. This is central to:

- Quantum chemistry,
- Condensed matter physics,
- Drug discovery and materials science.

There are several major approaches:

1. Trotter-Suzuki Decomposition

Break the time evolution into small slices:

$$e^{-iHt} \approx (e^{-iH_1 \Delta t} e^{-iH_2 \Delta t} \dots)^r$$

This is efficient when H can be decomposed into simple, non-commuting parts.

2. Variational Quantum Simulation

Use parametrized circuits and classical optimization to simulate ground states or dynamic behavior (e.g., Variational Quantum Eigensolver).

3. Linear Combination of Unitaries (LCU)

Decompose Hamiltonians into weighted sums of unitary operations, enabling controlled evolution with ancillary qubits.

4. Quantum Signal Processing

Advanced technique using polynomial approximations of time evolution, highly efficient for fault-tolerant quantum computers.

Challenges in Quantum Simulation

- **Gate depth**: Simulation algorithms often require deep circuits, which can be problematic for NISQ devices.
- **Error sensitivity**: Precise phase and amplitude dynamics must be preserved, increasing the need for error correction.
- **Encoding Hamiltonians**: Translating real-world problems into qubit Hamiltonians is often nontrivial.

Despite these hurdles, simulation remains one of the **most promising near-term applications** for quantum computing.

Theoretical Significance

Quantum walks and simulation algorithms reveal that quantum computers aren't just fast—they're **dynamically different**. They evolve through unitary operators rather than stochastic updates, enabling **non-classical mixing** and problem-specific interference.

These methods connect quantum computing not only with computer science but also deeply with physics, network theory, and systems engineering.

Conclusion: Walking the Quantum Path

Quantum walks and simulation algorithms are more than techniques—they are **quantum-native frameworks** for navigating and modeling complex spaces.

They let us explore enormous combinatorial or physical landscapes **coherently**, searching and evolving in ways classical algorithms can't.

As we build better quantum devices, these methods will lead the way in materials discovery, graph analysis, and fundamental science—proving once again that quantum computing is not about doing old things faster, but about doing entirely **new things**.

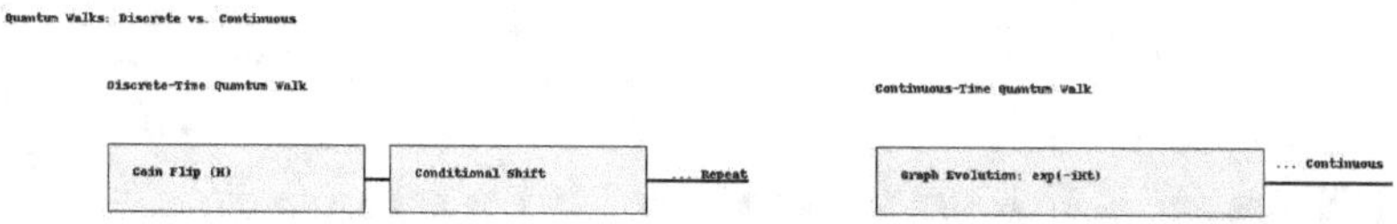

Discrete-Time Quantum Walk (DTQW):

- Requires coin and position registers.
- Alternates between coin flip and shift.
- Suitable for algorithmic search and decision problems.

Continuous-Time Quantum Walk (CTQW):

- Evolves directly via Hamiltonian (no coin needed).
- Models physical systems and network evolution.
- Preferred for simulation and energy dynamics.

The visual comparison of **Discrete-Time vs. Continuous-Time Quantum Walks**, including circuit elements and a summary of their features.

Section 5: Quantum Machine Learning

Chapter 25

Quantum Data and Feature Encoding

In classical computing, data is king. In quantum computing, **how** we encode that data—into amplitudes, qubits, and entanglement—is just as important as the data itself.

Quantum data encoding is the gateway between classical inputs and quantum processing. Without it, quantum computers would be empty shells with nothing to compute. With it, they become powerful engines capable of leveraging interference and entanglement to process information in fundamentally new ways.

As quantum computing matures—particularly in **quantum machine learning (QML)** and **hybrid quantum-classical systems**—feature encoding strategies are shaping up to be one of the **defining bottlenecks** and **key design choices**.

What Is Quantum Data?

There are two types of data in quantum computing:

1. **Classical data**: Structured or unstructured information from classical sources (e.g., images, time series, database entries) that must be encoded into quantum states.

2. **Quantum-native data**: Already exists as quantum states—e.g., quantum wave functions, entangled photons, or Hamiltonians.

This chapter focuses on encoding **classical data into quantum systems**—a prerequisite for almost all quantum machine learning, optimization, and simulation tasks.

Why Encoding Matters

Quantum computers cannot directly process classical data. The data must be mapped to quantum states, such as:

$$|\psi(x)\rangle = U(x)|0\rangle$$

Where

x

x is the input data and U(x) is a unitary transformation that embeds it into a high-dimensional Hilbert space.

How this encoding is performed affects:

- The **efficiency** of the algorithm,
- The **expressivity** of the quantum model,
- The **hardware compatibility**,
- The **ability to generalize** or learn patterns.

Popular Quantum Encoding Methods

1. Basis Encoding

Each classical bit is directly mapped to a qubit state.

- Input: binary vector `[1, 0, 1]`
- Output: $|101\rangle$
- Simple and efficient for binary data
- **Limitation**: No amplitude or phase encoding—can't exploit full quantum capabilities.

2. Amplitude Encoding

The entire input vector is embedded into the **amplitudes** of a quantum state:

$x=[x1\ ,x2\ ,...,xN\ ]\rightarrow|\psi\rangle=i=1\Sigma N\ xi\ |i\rangle$

- Efficient in terms of qubits: needs log(N) qubits for N-dimensional data
- Enables powerful parallelism
- **Limitation**: State preparation is complex and often inefficient on real devices.

3. Angle Encoding (Rotation Encoding)

Each feature is encoded as the angle of a quantum gate (typically a rotation gate like RX, RY, or RZ).

- Input: `[x1, x2, ..., xn]`

Encoding: RY(xi) applied to qubit

- Easy to implement on near-term devices
- Can be repeated in layered architectures for greater expressivity

4. QSample Encoding

Used for probability distributions: encodes a vector of probabilities into a quantum state using stochastic processes.

- Common in generative models and QML
- Useful in Bayesian methods and probabilistic reasoning

Choosing the Right Encoding Scheme

The best encoding depends on the **nature of the data**, **hardware constraints**, and **the type of algorithm** being used.

Encoding	Qubit Efficiency	Hardware Friendly	Expressive Power	Suitable For
Basis	High	Yes	Low	Binary data
Amplitude	Very high	No	Very high	Complex datasets
Angle (RY/RZ)	Medium	Yes	Moderate–High	Hybrid QML
QSample	Depends	Medium	High	Probabilistic models

Encoding in Quantum Circuits

In most quantum algorithms, data encoding is the **first step** in the circuit. It looks like:

```
|0⟩ —[Encoding(x1)]—■——————————
|0⟩ —[Encoding(x2)]—■— Algorithm
|0⟩ —[Encoding(x3)]—■——————————
```

The encoding block creates a **parameterized quantum state**, often followed by **entanglement gates** and **measurement**.

In **variational quantum circuits**, encoding layers are often repeated, interleaved with trainable layers.

Challenges in Quantum Encoding

- **Scalability**: Encoding large datasets requires either many qubits or complex gate sequences.
- **Noise Sensitivity**: Encoding steps increase circuit depth, which amplifies error.
- **Data Barren Plateaus**: Poor encoding can lead to flat optimization landscapes, especially in variational QML.
- **Noisy Oracles**: When encoding data indirectly via oracles or simulators, decoherence and approximation errors can dominate.

Encoding and Kernel Methods

In **quantum kernel methods**, the encoding defines a feature map

$\phi(x)$ such that:

$K(x ,x')=|\langle\phi(x)|\phi(x')\rangle|2$

This implicit kernel powers **quantum support vector machines**, **classification tasks**, and **clustering**. The choice of encoding **is the kernel**—making it the most critical design decision.

Encoding for Hybrid Algorithms

In **hybrid quantum-classical models**, classical optimizers depend heavily on the richness of encoded data. If encoding lacks expressivity or injectivity, no amount of training will recover performance.

To mitigate this:

- Use **multi-layer encodings**
- Combine **multiple encoding methods**
- Leverage **data re-uploading** techniques for richer representation

Conclusion: Encoding is the Interface

Quantum data encoding is the **interface** between classical data and quantum power.

It defines how the quantum system "sees" the problem and determines whether interference and entanglement can be harnessed meaningfully. As quantum algorithms become more complex, and data becomes more diverse, encoding strategies will increasingly define the **performance, feasibility, and interpretability** of quantum systems.

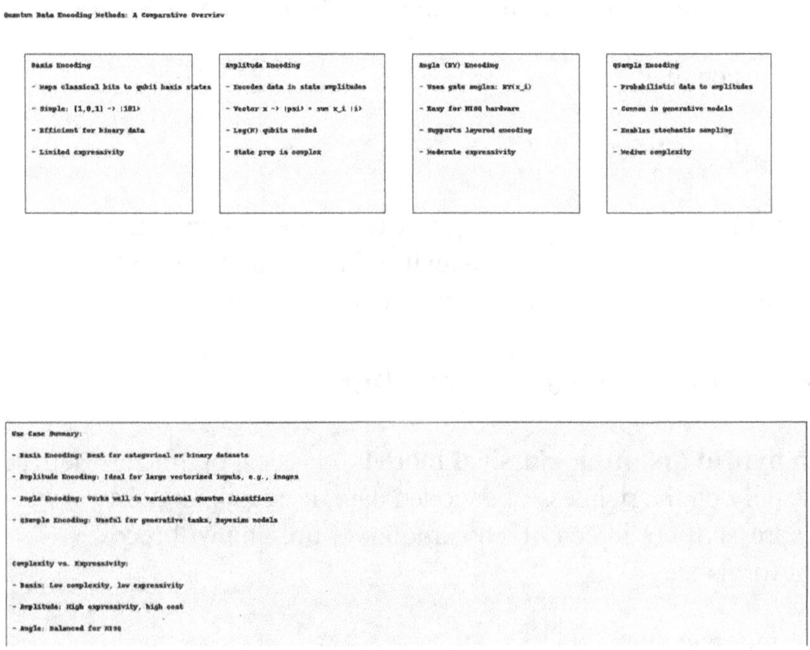

In quantum computing, **how you ask the question** matters. Encoding is how you ask.

The infographic comparing **quantum data encoding methods**—including Basis, Amplitude, Angle, and QSample encoding.

Chapter 26:

Quantum Neural Networks

The remarkable success of classical neural networks—from image recognition to large language models—has prompted researchers to ask: **Can we bring similar capabilities to quantum computers?** Enter **Quantum Neural Networks (QNNs)**—a class of quantum machine learning models inspired by the architecture and function of classical neural networks, yet fundamentally shaped by quantum mechanics.

Quantum Neural Networks aim to harness **superposition**, **entanglement**, and **quantum interference** to build models that can **learn**, **generalize**, and **adapt** in ways that could outperform classical networks, especially for data embedded in high-dimensional quantum spaces.

What Are Quantum Neural Networks?

Quantum Neural Networks are parameterized quantum circuits (PQCs) that mimic the layer-wise structure of classical neural nets. However, instead of real-valued matrix multiplications and nonlinear activation functions, QNNs operate with:

- **Quantum gates as weights**
- **Qubits as neurons**
- **Measurements as outputs**

Their parameters (e.g., angles of rotation gates) are trained using **classical optimization algorithms**—making them a central part of **hybrid quantum-classical computing**.

QNN Architecture Overview

A typical QNN consists of:

1. **Encoding Layer**
   Converts classical inputs into quantum states using angle or amplitude encoding.

2. **Parameterized Layers**
   Layers of quantum gates (RY, RZ, CX, etc.) with trainable parameters. These gates form the "weights" of the network.

3. **Entanglement Blocks**
   Introduce interactions between qubits, analogous to weight sharing or dense connections in classical nets.

4. **Measurement Layer**
   Measures qubits to generate output probabilities, often passed to a classical decision rule (e.g., softmax or threshold).

Popular QNN Models

1. Variational Quantum Classifier (VQC)

One of the most widely used QNN forms. Involves an encoding circuit followed by trainable layers. It's used for classification tasks and often integrated with SciKit-Learn or PyTorch backends.

2. Quantum Convolutional Neural Networks (QCNNs)

Inspired by classical CNNs. QCNNs apply quantum gates locally to small qubit neighborhoods and downsample via measurements. They are promising for data compression and pattern detection.

3. Quantum Boltzmann Machines (QBM)

A quantum analogue of classical Boltzmann Machines. It uses quantum probability distributions to model complex patterns, often via annealing or Hamiltonian sampling.

4. Quantum Recurrent Models

Still experimental, these attempt to model temporal or sequential data using iterative circuits with memory-like behavior.

Training a QNN

QNNs are trained using **variational methods**:

- Define a **loss function** (e.g., cross-entropy).
- Use a **classical optimizer** (e.g., gradient descent, Adam, COBYLA).
- Evaluate gradients via **parameter shift rules** or **finite differences**.
- Iterate to minimize loss over training data.

This forms a **hybrid loop** between the quantum processor (forward pass) and classical computer (back propagation and update).

Benefits of QNNs

- **Exponential Hilbert Space**: A small number of qubits can represent an exponentially large feature space.
- **Quantum Entanglement**: Enables richer correlations across inputs.
- **Potential for Speedup**: Certain classification problems may be solved more efficiently.

Challenges and Limitations

- **Barren Plateaus**: Large QNNs may have flat loss landscapes, making training slow or impossible.
- **Noisy Hardware**: Near-term quantum processors introduce decoherence and readout errors.
- **Data Bottleneck**: Encoding classical data remains a critical hurdle.
- **Scalability**: Limited qubit count restricts depth and complexity of models.

Applications of Quantum Neural Networks

- **Image and Signal Classification**
- **Fraud Detection and Financial Prediction**
- **Quantum Phase Recognition in Materials**
- **Anomaly Detection in Quantum Sensors**
- **Hybrid Quantum NLP Models**

Some of these are exploratory, while others are already being tested in quantum cloud environments (IBM Qiskit, PennyLane, TensorFlow Quantum, etc.).

QNNs vs Classical Neural Networks

Feature	Classical NN	Quantum Neural Network (QNN)
Input Space	Real-valued vectors	Encoded into quantum states
Layers	Linear + non-linear (e.g., ReLU)	Unitary gates + entanglement
Computation	Matrix ops	Quantum circuit evolution
Training	Backpropagation	Hybrid variational optimization
Potential Speedup	None (deterministic)	Probabilistic + quantum parallelism
Deployment	Mature on CPUs/GPUs	Emerging on quantum hardware

Future Directions

As QNN architectures mature, we expect to see:

- **QNN AutoML**: Automatic search over encoding and topology.
- **Noise-aware QNNs**: Circuits adapted to hardware-specific error models.
- **Quantum transformers** and **attention mechanisms**.
- Deeper integration into **edge computing**, **cloud AI pipelines**, and **quantum NLP**.

Conclusion: The Learning Mind of Quantum Machines

Quantum Neural Networks blend the adaptability of machine learning with the parallelism of quantum mechanics. While still in early stages, they represent a promising path forward for building quantum systems that **learn**, **adapt**, and **scale** to real-world tasks.

In the future, QNNs may unlock capabilities not just in quantum supremacy, but in **quantum intelligence**—computational models that learn from entangled experience and operate beyond classical imagination.

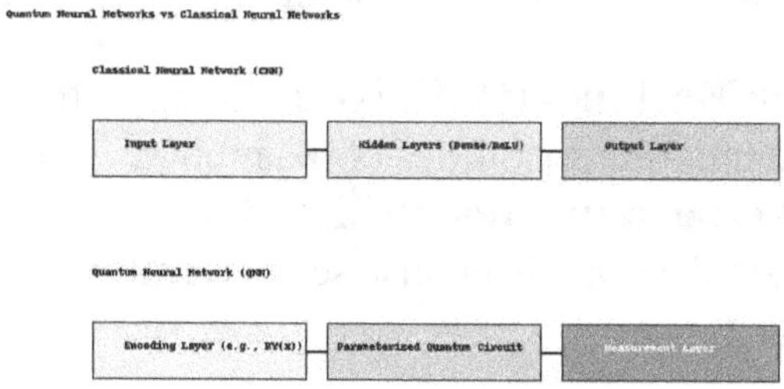

Key Differences:

- Classical NN: Layers apply matrix multiplications and activation functions.
- QNN: Uses quantum gates, entanglement, and measurement instead of weights.
- Training is hybrid: Quantum forward pass + classical optimization.
- QNNs can encode high-dim[illegible] parameters.

Use Cases:

- Classical NN: Vision, NLP, forecasting.
- QNN: Quantum chemistry, QML classification, anomaly detection, physics-aware models.

Chapter 27

Quantum Support Vector Machines (QSVMs)

Support Vector Machines (SVMs) are powerful classical machine learning models that find the optimal hyperplane to separate data into classes. But what happens when the data is too complex or too high-dimensional for classical algorithms to separate efficiently?

Quantum Support Vector Machines (QSVMs) aim to solve this problem by leveraging the unique capabilities of quantum computers—particularly **kernel estimation in high-dimensional Hilbert spaces**—to separate data that is otherwise inseparable using classical methods.

The Classical SVM Recap

A classical SVM works by:

- Finding a **hyperplane** that maximizes the margin between classes.
- Mapping data into a **higher-dimensional space** using a **kernel function** K(x ,x′).

  Solving a convex optimization problem to classify new points based on their relation to the margin.

The effectiveness of the SVM often comes from the choice of the kernel, which implicitly defines the shape of the decision boundary.

What Makes QSVM Different?

QSVMs use a **quantum computer to compute the kernel function**. The key idea is that quantum circuits can:

- Map classical data into **quantum states** $|\phi(x)\rangle$,
- Estimate the **inner product** between quantum states: $K(x, x')=|\langle\phi(x)|\phi(x')\rangle|2$,
- Provide access to a feature space that is **exponentially large**, depending on the number of qubits.

Instead of computing the kernel classically, a QSVM runs **quantum circuits** to estimate the overlap between encoded quantum states.

QSVM Architecture Overview

1. **Data Encoding Circuit**
   Encode classical input vectors into quantum states using a feature map

   U(x). This often involves parameterized RY or RZ rotations and entanglement gates.

2. **Quantum Kernel Estimation**
   For each pair (x ,x′), estimate the fidelity:

3. $K(x, x')=|\langle 0|U^\dagger(x')U(x)|0\rangle|2$

4. **Classical Optimization**
   Plug the quantum-computed kernel matrix into a **classical SVM optimizer** (e.g., from SciKit-Learn) to solve the optimization problem.

5. **Prediction Phase**
   Use the trained model to classify new inputs by repeating the quantum kernel estimation.

Feature Maps in QSVM

Choosing the right **feature map** is critical. Some popular options include:

- **ZZFeatureMap**: Involves RZ rotations followed by entangling CZ gates.
- **Pauli Feature Maps**: Use Pauli-based operators for encoding data.
- **Custom Maps**: Designed to encode specific domain knowledge (e.g., graphs, molecules).

The choice determines the **geometry of the quantum feature space** and the **complexity of the decision boundary**.

Advantages of QSVM

- **High-dimensional expressivity**: Access exponentially large feature spaces with modest qubit counts.
- **Implicit non-linearity**: Naturally encodes complex relationships via quantum interference.
- **Potential quantum advantage**: Can achieve separation where classical SVMs fail due to kernel limitations.
- **NISQ-compatible**: Requires only shallow circuits and no training inside the quantum loop.

Limitations and Considerations

- **Noise sensitivity**: Kernel estimation requires high-fidelity quantum circuits.

- **Scalability**: Pairwise comparisons between data points scale quadratically.
- **Feature map design**: Poorly chosen encodings can collapse expressivity or lead to barren plateaus.
- **Hardware availability**: Still limited by current quantum gate fidelity and qubit counts.

Real-World Applications

QSVMs have been applied or proposed for:

- **Financial anomaly detection**
- **Drug molecule classification**
- **Image recognition (e.g., MNIST subsets)**
- **Fraud and cybersecurity analysis**
- **Quantum chemistry pattern recognition**

Several quantum cloud platforms (e.g., IBM Qiskit, PennyLane) already offer QSVM templates for experimentation.

QSVM vs Classical SVM

Feature	Classical SVM	Quantum SVM (QSVM)
Kernel Space	Predefined (linear, RBF, poly)	Quantum state overlap

Feature Mapping	Explicit, fixed	Implicit via quantum circuits
Optimization	Classical	Classical
Computational Cost	Polynomial	High pre-processing (quantum kernel)
Potential Advantage	Limited to kernel choice	Access to larger feature spaces

Conclusion: Quantum Kernels as a Gateway to QML

QSVMs offer a pragmatic bridge between **quantum computing and traditional machine learning**. By using quantum processors only for what they're uniquely good at—**high-dimensional kernel estimation**—QSVMs make quantum advantage accessible even on today's NISQ devices.

They represent the first class of **hybrid quantum models** that are genuinely competitive with classical baselines on small datasets, and they provide critical insight into how quantum learning might evolve.

In the future, as quantum hardware improves and kernel estimation scales, QSVMs may become foundational tools for **quantum-native learning**, helping solve complex problems that defy classical boundaries.

Chapter 28

Variational Quantum Algorithms (VQAs)

In the NISQ era—where quantum computers are powerful yet noisy and limited—**Variational Quantum Algorithms (VQAs)** have emerged as the most promising pathway for practical quantum computing.

By blending **quantum circuits** with **classical optimization loops**, VQAs provide a framework to tackle real-world problems in chemistry, optimization, and machine learning using today's hardware capabilities.

What Are VQAs?

VQAs are **hybrid quantum-classical algorithms**. They use:

- A **parameterized quantum circuit** (also called an ansatz) to process information,
- A **classical optimizer** to tune the parameters based on a cost function.

The quantum part prepares and evolves quantum states; the classical part interprets the results and adjusts the parameters to improve performance.

General Workflow of a VQA

1. **Initialize Parameters**
   Randomly or heuristically choose initial values for rotation angles or gate parameters.

2. **Prepare Quantum Circuit (Ansatz)**
   Construct a parameterized quantum circuit using gates like RY, RX, and entangling gates (e.g., CNOT).

3. **Measure Observables**
   Measure quantum states to evaluate a **cost function**, such as energy, distance, or loss.

4. **Classical Optimization**
   Use a classical algorithm (e.g., gradient descent, SPSA, COBYLA) to update parameters.

5. **Iterate**
   Repeat steps 2–4 until convergence.

This hybrid loop continues until the system finds a minimum in the cost landscape.

Why Use a Variational Approach?

Quantum computers excel at preparing **complex superpositions** and evaluating observables, but they struggle with depth, noise, and scalability. The **variational method delegates the hard optimization problem to classical machines**, reducing quantum circuit complexity.

This makes VQAs well-suited for current devices with:

- Short coherence times,
- Limited qubit counts,
- Restricted connectivity.

Popular Types of VQAs

1. Variational Quantum Eigensolver (VQE)

Used in quantum chemistry to estimate the ground state energy of molecules.

- Cost function: Expected value of a Hamiltonian H,
- Ansatz: Problem-specific or hardware-efficient circuits,
- Application: Hydrogen molecule, LiH, BeH_2, etc.

2. Quantum Approximate Optimization Algorithm (QAOA)

Used to solve combinatorial optimization problems (e.g., MaxCut, SAT).

- Alternates between two Hamiltonians: cost and mixing,
- Parameters represent durations or strengths,
- Application: Graph optimization, scheduling, portfolio design.

3. Variational Quantum Classifier (VQC)

Used in supervised learning. Inputs are encoded into quantum states and classified by variational circuits.

- Output: Predicted class label,
- Optimized using cross-entropy or hinge loss.

Key Components of a VQA

- **Ansatz Design**: The choice of parameterized circuit. Can be hardware-efficient or problem-informed.
- **Cost Function**: Observable that reflects the quality of the solution.
- **Classical Optimizer**: Minimizes the cost using parameter updates (e.g., Adam, Nelder-Mead, SPSA).
- **Measurement Strategy**: Determines how to estimate observables with as few shots as possible.

Strengths of VQAs

- **Hardware-friendly**: Shallow circuits and local gates work well on real quantum hardware.
- **Flexible**: Can be tailored for chemistry, optimization, ML, and simulation.
- **Error-tolerant**: Naturally robust to some forms of noise.
- **Scalable (theoretically)**: Amenable to modular extension as hardware improves.

Limitations and Challenges

- **Barren Plateaus**: Flat regions in the cost landscape make training difficult.
- **Noise Accumulation**: Still affected by decoherence, especially in larger circuits.
- **Ansatz Engineering**: Poorly chosen ansatz can bottleneck performance.

- **Optimizer Instability**: Classical optimizers may be sensitive to shot noise or hyper parameters.

Application Areas

- **Quantum Chemistry**: Estimate molecular energies (VQE).
- **Optimization**: Solve real-world constrained problems (QAOA).
- **Machine Learning**: Build trainable quantum classifiers (VQC).
- **Finance**: Risk minimization, asset pricing, portfolio optimization.
- **Physics**: Simulate many-body systems and condensed matter phenomena.

VQAs in Practice

Platforms like **Qiskit**, **PennyLane**, and **TensorFlow Quantum** offer toolkits for building and training VQAs. Users can define custom ansatzes, plug in classical optimizers, and test on actual quantum hardware or simulators.

Example (Qiskit-style):

```
ansatz = TwoLocal(num_qubits=4, rotation_blocks='ry',
entanglement_blocks='cz')

optimizer = COBYLA()

vqe = VQE(ansatz=ansatz, optimizer=optimizer,
quantum_instance=backend)
```

Conclusion: Variational Thinking for Quantum Reality

VQAs are the workhorses of today's quantum computing era. They offer a practical compromise—using quantum systems for what they do best (state preparation and evaluation) and classical systems for what they do best (optimization and control).

As quantum hardware evolves, VQAs will serve as the foundation for increasingly powerful hybrid applications—paving the way to quantum-enhanced solutions in chemistry, logistics, AI, and beyond.

The infographic illustrating the **Variational Quantum Algorithm (VQA) workflow** and a comparison between **VQE**, **QAOA**, and **VQC**

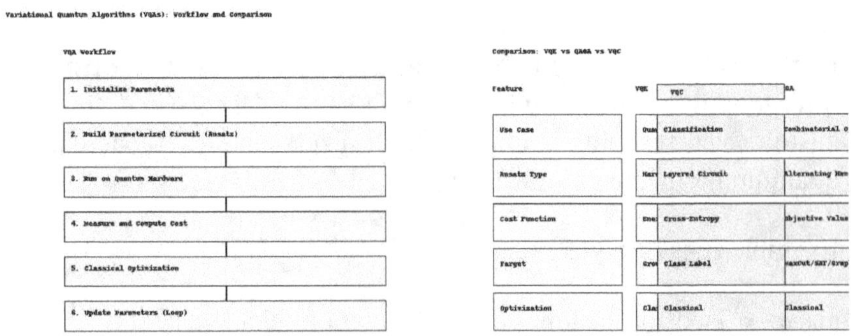

Chapter 29

Quantum Reinforcement Learning

In traditional machine learning, **reinforcement learning (RL)** is the framework where agents learn optimal actions by interacting with an environment and receiving feedback in the form of rewards. In the **quantum version of RL**, we extend this paradigm by introducing **quantum states, operations, and memory** into the learning loop.

Quantum Reinforcement Learning (QRL) is a hybrid frontier—where classical and quantum resources are combined to create learning systems that potentially outperform their classical counterparts in adaptability, representation power, or convergence speed.

What Is Reinforcement Learning (Brief Recap)?

Reinforcement learning is based on the agent-environment loop:

- **Agent**: Makes decisions based on current observations.
- **Environment**: Responds with new states and rewards.
- **Goal**: Maximize the cumulative reward over time.

Mathematically, this process is modeled as a **Markov Decision Process (MDP)**, where transitions depend on the agent's actions and the current state.

What Is Quantum Reinforcement Learning?

Quantum Reinforcement Learning introduces **quantum elements** into RL in one or more of the following areas:

1. **Quantum State Representation**: States and actions are encoded in quantum registers.
2. **Quantum Policy Models**: The decision-making model (policy or value function) is implemented via parameterized quantum circuits.
3. **Quantum Environment Simulation**: Environments (especially in physics or chemistry) are simulated on quantum devices.
4. **Quantum Memory**: Agents retain information using quantum systems (e.g., quantum recurrent circuits).

QRL aims to **improve exploration**, **reduce sample complexity**, and **accelerate learning** by leveraging quantum superposition, entanglement, and interference.

Types of QRL Frameworks

1. Quantum Policy Gradient Methods

The policy (agent behavior) is modeled as a quantum circuit with tunable parameters. The agent samples actions based on quantum measurements and updates parameters using classical optimizers.

- Similar to classical policy gradient methods.
- Encoding input states via RY/RZ rotations.
- Gradient estimated via **parameter shift rule**.

2. Quantum Value-Based Methods (Q-Q Learning)

A quantum circuit estimates the value of states or state-action pairs. Updates resemble classical Q-learning but use quantum models to represent Q-functions.

- Hybrid setups: quantum for value function, classical for policy.
- Advantage: Q-function benefits from quantum representation of state space.

3. Quantum Environment Models

The environment is simulated on a quantum computer—especially useful in **quantum control**, **quantum chemistry**, and **quantum materials**. The agent can then learn policies to manipulate quantum systems more efficiently.

Benefits of Quantum RL

- **Efficient Exploration**: Superposition allows simultaneous evaluation of multiple state-action paths.
- **Quantum Memory**: Qubits can encode and retain richer historical information.
- **Scalable Representation**: High-dimensional states and action spaces are compressed using qubit-efficient encodings.
- **Speedup**: Potential speedups in updating value or policy gradients.

Challenges and Limitations

- **Noise and Decoherence**: Quantum circuits are error-prone, limiting depth and stability.

- **Gradient Estimation**: High variance and barren plateaus can hinder training.
- **Sample Inefficiency**: Despite theoretical speedups, practical QRL may require many episodes to converge.
- **Benchmarking**: Lack of standard environments and performance metrics in QRL.

Applications of QRL

- **Quantum Control Tasks**: Learning control pulses for quantum gates or operations.
- **Autonomous Quantum Error Correction**: Agents learn to detect and correct errors.
- **Financial Trading Agents**: Combining quantum optimization with market interaction.
- **Quantum Game AI**: Agents navigating quantum environments or multi-agent quantum games.
- **Quantum Robotics (theoretical)**: Decision-making for quantum-augmented robotic systems.

Example: Quantum Policy Gradient Loop

1. Encode current state into quantum circuit.
2. Apply parameterized quantum circuit to generate policy.
3. Sample action from quantum measurement.
4. Execute action in environment.
5. Receive reward and next state.

6. Update parameters using classical optimizer.

This forms a **hybrid loop**, much like in VQAs, where quantum circuits are embedded in RL agents.

Feature	Classical RL	Quantum RL
State Representation	Real-valued vectors	Quantum states or amplitudes
Policy/Value Model	Neural networks	Parameterized quantum circuits
Exploration	Stochastic sampling	Superposition + entanglement
Training Loop	Classical	Hybrid (quantum + classical)
Environment	Simulated or real-world	Can include quantum environments

Future Outlook

QRL is still in its infancy but is rapidly developing. Expected trends include:

- **Standardized QRL Benchmarks** (quantum OpenAI Gym equivalents),
- **Quantum Multi-Agent RL** systems,
- **Noise-resilient QRL architectures**,
- Applications in **quantum chemistry**, **materials science**, and **automated quantum hardware calibration**.

Conclusion: Learning in Quantum Landscapes

Quantum Reinforcement Learning brings adaptive learning to the quantum domain. It's not just about speeding up RL, but about rethinking how an **agent interacts with and learns from quantum systems**.

As quantum processors improve and hybrid models become more sophisticated, QRL may become a critical tool in building intelligent quantum agents—agents capable of making decisions not only **about** quantum systems but **within** them.

The diagram illustrating the **Quantum Reinforcement Learning (QRL)** agent-environment loop—compared side-by-side with classical RL.

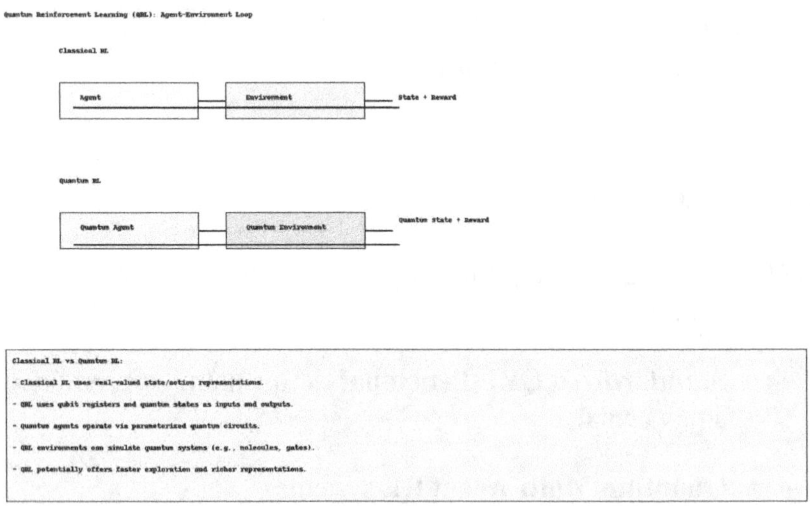

Chapter 30

Hybrid Quantum-Classical AI Systems

As quantum computers continue to evolve, the vision of a fully quantum artificial intelligence (AI) remains compelling—but currently out of reach. However, we don't need to wait for universal fault-tolerant quantum computers to start building powerful AI systems. Instead, we can combine the **best of both worlds** by fusing classical AI techniques with quantum computing.

Hybrid quantum-classical AI systems are emerging as one of the most promising near-term pathways to quantum advantage. These systems harness **quantum resources for specific subroutines** (like optimization, kernel estimation, or encoding), while leveraging the **maturity and scalability of classical machine learning** infrastructure.

What Is a Hybrid Quantum-Classical AI System?

A hybrid system combines quantum and classical components in a **feedback loop**, where classical AI models handle tasks like preprocessing, training loops, or optimization, while quantum circuits are used for:

- Feature embedding,
- Inference,
- Subspace sampling,
- Specialized function evaluation (e.g., kernel, Hamiltonian expectation).

The two systems **cooperate dynamically**, exchanging data and gradients to build a unified AI pipeline.

Why Hybrid?

Quantum computers are still:

- Noisy,
- Resource-constrained,
- Not general-purpose.

But they **excel** at tasks involving:

- Exponentially large vector spaces,
- Hard-to-classically-simulate operations (e.g., entangled evolution),
- Sampling from complex distributions.

Hybrid systems allow us to plug quantum modules into existing AI pipelines, extracting benefit **without total dependence on quantum hardware**.

Core Architectures of Hybrid Systems

1. Quantum-Enhanced Feature Embedding

Use quantum circuits to encode classical data into high-dimensional Hilbert space. Output is fed into a classical model for classification or regression.

Use case: Quantum kernel methods, hybrid SVMs, quantum CNN front-ends.

2. Variational Circuits as Layers

Parameterize quantum circuits as "layers" within deep learning models. Their outputs become part of classical forward/backward propagation.

Use case: TensorFlow Quantum, PennyLane + PyTorch, quantum-enhanced autoencoders.

3. Hybrid Optimizers

Quantum computers evaluate sub functions (e.g., expected value of cost Hamiltonians), while classical optimizers drive convergence.

Use case: VQE, QAOA, Quantum GANs.

4. Quantum Inference Modules

Use quantum processors for sampling, generative inference, or probabilistic updates.

Use case: Quantum Boltzmann Machines, quantum variational inference.

Design Patterns in Hybrid Systems

Component	Role	Implementation
Data Preprocessor	Normalize or structure input	Classical (e.g., NumPy)
Encoder	Map data to quantum circuit	Quantum (Qiskit, PennyLane)
Model	Process data & produce output	Hybrid (quantum circuit + classical layers)

| Optimizer | Tune model parameters | Classical (e.g., Adam, SPSA) |

| Loss Evaluator | Assess performance | Classical (loss function) |

Benefits of Hybrid Systems

- **Practical**: Works with current quantum hardware.
- **Flexible**: Swap in quantum modules based on task need.
- **Composable**: Integrates with established ML toolchains.
- **Explainable**: Quantum layers may offer new interpretability angles.
- **Performance Boost**: For problems with complex landscape, quantum subspaces may help convergence or expressivity.

Challenges of Hybrid Integration

- **Latency**: Quantum-classical communication can be slow.
- **Noise**: Quantum operations introduce variability in outputs.
- **Gradient Calculation**: Backpropagation through quantum circuits is non-trivial and must use specialized techniques (e.g., parameter-shift rule).
- **Data Bottleneck**: Encoding large classical datasets into quantum states remains a challenge.

Software Ecosystem for Hybrid AI

- **PennyLane** – Integrates with PyTorch and TensorFlow; supports automatic differentiation through quantum circuits.
- **TensorFlow Quantum** – Google's quantum ML framework using Cirq.
- **Qiskit Machine Learning** – Provides QSVM, VQC, and feature map support.
- **MindQuantum** – From Huawei, focuses on VQAs and hybrid modeling.

These frameworks abstract much of the quantum-classical interaction, making it easier to prototype models.

Applications of Hybrid Systems

- **Drug Discovery**: Quantum-enhanced models to screen molecules faster.
- **Financial Modeling**: Portfolio optimization using QAOA within AI systems.
- **Image Classification**: Quantum preprocessing + classical CNNs.
- **Materials Science**: Quantum simulations feeding into generative classical models.
- **Quantum NLP**: Embedding symbolic structures into quantum Hilbert space.

Example Workflow: Quantum-Enhanced Classifier

1. Classical system collects and cleans data.

2. Quantum circuit encodes features (e.g., via RY rotations + entanglement).
3. Output passed to classical neural network layer.
4. Prediction generated; loss calculated.
5. Classical optimizer updates quantum circuit parameters.

This loop continues until convergence.

Outlook: AI in the Quantum Era

Hybrid quantum-classical systems are not a compromise—they are the **architecture of the near future**. As both quantum hardware and software evolve, these systems will become increasingly powerful, flexible, and widespread.

They don't replace classical AI—they **extend it**, offering tools that classical models could never access on their own.

In the next few years, expect to see hybrid AI systems move from research labs to real-world deployments—first in narrow, high-value domains, and eventually as **building blocks of quantum-native intelligence**.

The following architecture is a diagram of a **Hybrid Quantum-Classical AI System**, including labeled modules and data flow.

Hybrid Quantum-Classical AI System Architecture

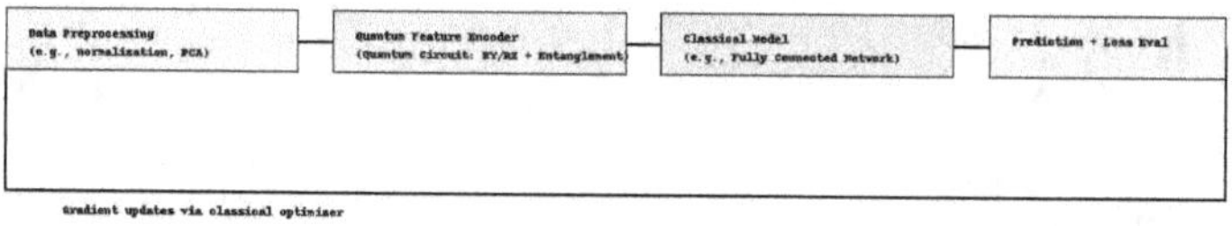

Key Components:

- Classical Preprocessing: Scales, filters, or compresses raw data.
- Quantum Encoder: Maps data into quantum feature space using qubit rotations.
- Classical Neural Net: Learns patterns from quantum-encoded outputs.
- Loss Module: Evaluates performance and guides optimization.
- Optimizer: Updates parameters in classical and quantum parts.

Benefits:

- Utilizes quantum resources only where impactful.
- Integrates with existing ML infrastructure (e.g., PyTorch, TensorFlow).
- Scalable and modular for future quantum upgrades.

Section 6: Quantum Cryptography and Security

Chapter 31

Quantum Key Distribution (QKD)

In an era where cybersecurity is more critical than ever, **Quantum Key Distribution (QKD)** offers a revolutionary promise: communication security rooted not in computational difficulty—but in the immutable laws of physics.

Unlike classical encryption methods (which can be broken with enough time or a powerful enough computer), QKD ensures that any **eavesdropping attempt is detectable**. This is possible thanks to the **fundamental properties of quantum mechanics**, such as the no-cloning theorem and measurement-induced disturbance.

QKD is the first real-world application of quantum information science to be **commercially deployed**—a crucial stepping stone toward the broader quantum internet.

The Core Idea of QKD

QKD allows two parties—traditionally called **Alice** and **Bob**—to generate a **shared secret key** over an insecure channel, using quantum states. If an eavesdropper (**Eve**) tries to intercept the quantum communication, she inevitably introduces **disturbances**, which can be detected.

The key generated can then be used with **classical encryption protocols** (e.g., one-time pad, AES) to enable **unbreakable encryption**.

BB84: The First and Most Famous Protocol

The **BB84 protocol**, proposed by Charles Bennett and Gilles Brassard in 1984, is the most well-known and widely implemented QKD protocol.

BB84 works as follows:

1. **Quantum Transmission**
   Alice sends qubits to Bob, each prepared in one of four polarizations (e.g., 0°, 90°, 45°, 135°), chosen at random.

2. **Measurement**
   Bob measures each qubit in a randomly chosen basis (rectilinear or diagonal).

3. **Public Discussion**
   Alice and Bob compare basis choices publicly (not measurement results), and keep only bits where they used the same basis.

4. **Error Checking**
   They compare a subset of bits to estimate error rate. If it's below a threshold, they proceed.

5. **Key Extraction**
   They apply classical algorithms for **error correction** and **privacy amplification** to finalize the shared secret key.

If Eve tries to measure the qubits en route, she will disturb them due to the **quantum measurement principle**, causing detectable errors.

Why QKD Is Unique

- **Provable Security**: Based on physics, not assumptions about computational difficulty.

- **Eavesdropper Detection**: Any interception attempt changes quantum states.

- **No Cloning**: The no-cloning theorem forbids perfect copies of unknown qubits.

Variants of QKD

- **BB84** – The standard discrete-variable protocol.
- **E91** – Based on entangled photon pairs (proposed by Ekert in 1991).
- **Decoy-State QKD** – Introduces "decoy" signals to detect photon-number-splitting attacks.
- **Continuous Variable (CV-QKD)** – Uses quadrature measurements of light rather than discrete qubits.
- **Measurement-Device-Independent QKD (MDI-QKD)** – Immunizes against detector side-channel attacks by having both parties send quantum states to an untrusted third party.

QKD Network Architectures

QKD can be deployed in various network models:

- **Point-to-point fiber links** (e.g., city-to-city secure channels),
- **Satellite-based QKD** (e.g., China's Micius satellite),
- **Quantum repeaters** (for future long-range entanglement),
- **Metropolitan QKD networks** (e.g., secure financial sectors, data centers).

QKD in the Real World

- **China** has deployed a 2,000-km QKD backbone from Beijing to Shanghai.
- **Europe** is working on EuroQCI (Quantum Communication Infrastructure).
- **Commercial Vendors**: Toshiba, ID Quantique, and QuintessenceLabs offer QKD solutions.
- **Applications**: Government communications, financial institutions, defense, and secure voting.

Advantages of QKD

- Immune to future quantum computer attacks (unlike RSA or ECC).
- Provides forward secrecy even if classical data is recorded now.
- Can integrate with classical encryption protocols.

Limitations and Challenges

- **Distance limitations**: Photons degrade over long fibers.
- **Hardware costs**: Requires specialized photon sources, detectors, and low-loss infrastructure.
- **Speed**: Key generation rates are relatively low compared to classical key exchange.
- **Side-channel attacks**: Real-world imperfections may open attack vectors.

QKD vs Classical Cryptography

Feature	Classical Encryption (e.g., RSA)	Quantum Key Distribution (QKD
Security Basis	Mathematical assumptions	Physical laws (quantum mechanics)
Breakable by Quantum?	Yes (e.g., Shor's algorithm)	No (detects eavesdropping)
Requires Quantum HW?	No	Yes
Distance Limitation	No	Yes (fiber/satellite bound)
Practical Today?	Yes, widely used	Yes, in specific high-security use

Future Outlook: Toward the Quantum Internet

QKD is the first building block of the **quantum internet**—a future network that will transmit quantum states and entanglement across global distances.

- **Quantum repeaters** and **error-corrected photonic links** will eventually allow global QKD.
- Integration with **5G/6G infrastructure** is being explored.
- Advances in **quantum satellite tech** will make cross-continental quantum-secured communication feasible.

Conclusion: Securing the Future with Physics

Quantum Key Distribution shifts the paradigm of cryptography. It doesn't just **raise the bar** for attackers—it **changes the game**, making secure communication an inherent feature of the laws of nature.

As quantum threats to classical encryption emerge, QKD will play an increasingly central role in ensuring that **privacy, confidentiality, and trust** remain unbroken in the quantum age.

Chapter 32

Post-Quantum Cryptography (PQC)

The advent of quantum computing poses a serious threat to modern cryptographic systems. Algorithms like RSA, Diffie-Hellman, and Elliptic Curve Cryptography (ECC)—cornerstones of secure digital communication—can be **broken by a sufficiently powerful quantum computer** using **Shor's algorithm**.

But not all is lost. **Post-Quantum Cryptography (PQC)**—also called **Quantum-Resistant Cryptography**—is a field that develops **classical encryption algorithms** designed to resist quantum attacks. PQC offers a path forward that's both practical and deployable today, without waiting for quantum hardware.

Why We Need Post-Quantum Cryptography

Quantum algorithms threaten classical cryptography in two major ways:

- **Shor's Algorithm**: Efficiently factors large numbers and solves discrete logarithms. This breaks RSA, ECC, and Diffie-Hellman.
- **Grover's Algorithm**: Speeds up brute-force attacks by a square root, weakening symmetric encryption and hash functions.

While symmetric cryptography (e.g., AES-256) survives with minor changes, **public-key encryption is fundamentally vulnerable**. PQC provides new schemes that withstand these threats.

Core Characteristics of PQC Algorithms

PQC algorithms are:

- **Run on classical computers** (no quantum hardware needed),
- **Designed to resist attacks from quantum computers**,
- **Tested against known quantum algorithms**,
- Intended for **drop-in replacement** of RSA/ECC in today's infrastructure.

Main Families of PQC Algorithms

1. Lattice-Based Cryptography

- Based on the hardness of problems like **Learning With Errors (LWE)** or **Short Integer Solutions (SIS)**.
- Well-studied, efficient, and scalable.
- Examples: **Kyber**, **Dilithium**, **FrodoKEM**.

2. Code-Based Cryptography

- Based on decoding random linear codes (e.g., Goppa codes).
- Long-standing security record.
- Examples: **Classic McEliece**.

3. Multivariate Polynomial Cryptography

- Based on solving multivariate equations over finite fields.

- Compact and efficient signatures.
- Examples: **Rainbow** (now deprecated due to structural vulnerabilities).

4. Hash-Based Signatures

- Rely on the one-wayness of hash functions.
- Stateless or stateful variants.
- Examples: **SPHINCS+**, **XMSS**.

5. Isogeny-Based Cryptography

- Uses hard problems on elliptic curve isogenies.
- Promising for key exchange with small key sizes.
- Example: **SIKE** (though it was broken in 2022).

NIST PQC Standardization Effort

The U.S. National Institute of Standards and Technology (NIST) has been leading a **global competition** to standardize quantum-resistant algorithms.

- **Round 3 finalists (2022)**:
    - **Kyber** (key encapsulation) – Winner
    - **Dilithium** (digital signatures) – Winner
    - **SPHINCS+** – Winner for hash-based signatures

These algorithms are expected to replace RSA and ECC in TLS, VPNs, and secure messaging protocols in the near future.

Comparison: PQC vs Classical and Quantum Crypto

Feature	RSA / ECC	Post-Quantum Crypto	Quantum Key Distribution (QKD)
Runs on Classical HW	✅	✅	❌ (needs quantum HW)
Quantum-Secure	❌	✅	✅
Deployment Maturity	✅	In transition	Experimental / niche
Key Exchange Method	Math-based	Math-based (new problems)	Quantum states
Suitable for Internet	✅	✅	Limited (point-to-point)

Challenges in PQC Deployment

- **Larger key sizes**: Some PQC schemes (e.g., McEliece) require huge public keys.
- **Backward compatibility**: Legacy devices may struggle with new cryptosystems.
- **Performance tuning**: New algorithms may be slower than classical ones for small inputs.

- **Security proof maturity**: Although promising, some methods are newer and less battle-tested.

Future Outlook for PQC

- **Global migration efforts** will pick up after NIST finalizes standardization.
- Governments and enterprises will **dual-stack crypto** during the transition.
- PQC will become part of a **multi-layered security model**, often coexisting with quantum cryptography (QKD).
- Expect integration into **post-quantum messaging**, **digital identity**, and **blockchain platforms**.

Conclusion: Classical Protection for the Quantum Age

Post-Quantum Cryptography is our first line of defense in the quantum future. It offers a **practical**, **hardware-agnostic**, and **scalable** way to protect communications from the quantum threat —**today**.

Unlike QKD, it requires **no new infrastructure** and can be adopted incrementally. As we move closer to the era of quantum computing, PQC ensures that the systems we rely on remain **secure, stable, and private**, even in the face of technological upheaval.

Chapter 33

Quantum Attacks on RSA and ECC

Modern digital security—whether it's securing your email, online banking, or corporate VPN—relies heavily on public-key cryptography. Protocols like **RSA** and **Elliptic Curve Cryptography (ECC)** form the backbone of internet security, digital signatures, and encrypted communication.

But with the emergence of quantum computing, this foundation is under existential threat. Thanks to **Shor's algorithm**, a sufficiently powerful quantum computer can **efficiently break both RSA and ECC**, rendering decades of cryptographic infrastructure obsolete.

RSA and ECC: A Quick Refresher

RSA (Rivest–Shamir–Adleman)

- Based on the **factoring problem**: It's hard to factor a large number

  $N=p\times q$ into its prime components.

- Security depends on the computational difficulty of factoring large integers.
- Used in TLS, VPNs, digital signatures.
- **ECC (Elliptic Curve Cryptography)**
- Based on the **elliptic curve discrete logarithm problem (ECDLP)**: Given points P **and Q find k.**

- More efficient than RSA: smaller key sizes with equivalent security.
- Widely used in mobile devices, cryptocurrencies, messaging apps.

Enter Quantum Computing

Quantum computers operate fundamentally differently from classical machines. They exploit **superposition** and **entanglement** to process information in parallel at scale.

The biggest threat to public-key cryptography comes from one quantum algorithm:

Shor's Algorithm

- Developed by Peter Shor in 1994.
- Solves both integer factorization and discrete logarithms **exponentially faster** than the best classical algorithms.
- If implemented on a fault-tolerant quantum computer, it can break:
  - **RSA** by factoring N,
  - **ECC** by solving ECDLP.

Attack Complexity: Classical vs Quantum

Current quantum devices (e.g., IBM, IonQ, Rigetti) are still too small and noisy. But breaking RSA-2048 or ECC-256 would require:

- **Thousands to millions of logical qubits**,
- **Error correction layers** (e.g., surface codes),
- **Long coherence times**,
- **Deep gate fidelity**.

Estimates suggest that a **fully fault-tolerant quantum computer** could break RSA-2048 in a matter of hours using around **20 million physical qubits**, assuming 1% physical gate error rates and 1000 logical qubits.

Timeline and Risk

- While large-scale quantum computers aren't here yet, **data harvested today can be decrypted later**.
- This threat is called **"Harvest Now, Decrypt Later."**
- **Governments and enterprises** are urged to begin migrating to **post-quantum cryptography** before quantum computers are practical.

Impact Zones

Sector	At Risk With Quantum	Why It Matters
Web Security	Yes (HTTPS/TLS)	Breaks RSA key exchange
VPNs and Firewalls	Yes	Compromises authentication
Software Updates	Yes	Falsify signatures (code injection)

Sector	At Risk With Quantum	Why It Matters
Blockchain	Yes (ECC-based coins)	Signatures can be forged
Messaging (e.g., Signal)	Yes	ECC-based identity compromise
IoT Devices	Yes	Often hard-coded with ECC keys

Examples of Quantum Attack Scenarios

1. Intercepting TLS Communications

An attacker records encrypted web traffic today (RSA key exchange). Later, they decrypt it using a quantum computer by factoring the private key.

2. Forging Digital Signatures

Shor's algorithm allows forging of signatures in ECC and RSA-signed code, making malicious firmware or software updates appear legitimate.

3. Blockchain Takeover

Bitcoin and Ethereum (older wallets) use ECC. A quantum attacker could steal coins by deriving private keys from public ones exposed on the chain.

Mitigation Strategy

1. **Transition to Post-Quantum Cryptography**
   Use quantum-resistant algorithms like Kyber, Dilithium, or SPHINCS+.

2. **Use Hybrid Key Exchange**
   Combine classical and post-quantum methods to hedge against near-term threats.

3. **Shorten Key Lifetimes**
   Avoid long-lived keys and static certificates that can be harvested.

4. **Audit Cryptographic Inventory**
   Identify all uses of RSA/ECC across systems.

5. **Stay Updated on Standards**
   Follow NIST PQC developments and integrate standardized schemes.

Conclusion: A Cryptographic Turning Point

Quantum computers pose a real, existential threat to RSA and ECC. Although we're not there yet, **the timeline is uncertain**, and the implications are too severe to ignore.

By understanding how quantum algorithms like Shor's disrupt traditional cryptography, we prepare ourselves to **build forward-compatible, future-proof systems**—and ensure the continued privacy and security of digital information in a quantum world.

Chapter 34

Blind Quantum Computing (BQC)

As quantum computers become available through cloud services, concerns arise over the **privacy and security** of quantum computation. In classical computing, users often encrypt data before uploading it to the cloud—but **how do you encrypt quantum data**, or keep a quantum computation private?

Enter **Blind Quantum Computing (BQC)**, a novel cryptographic protocol that allows a client to **delegate quantum computation to a remote quantum server** while keeping both the **input, output, and the computation itself hidden**.

BQC is one of the most profound applications of **quantum cryptography**, combining principles from quantum entanglement, measurement-based quantum computation, and interactive proof systems.

What Is Blind Quantum Computing?

Blind Quantum Computing is a **delegated quantum computation protocol**. It allows a **client with limited quantum capabilities** (or even just classical control) to instruct a **remote quantum computer** to perform computations without revealing:

- The **input data**,
- The **algorithm or circuit** being run,
- The **final result**.

The **quantum server performs the computation**, but learns nothing about what it's computing.

Why BQC Matters

- **Privacy**: Protects sensitive quantum computations, such as proprietary simulations or encrypted operations.
- **Cloud Compatibility**: Enables secure use of third-party quantum services (e.g., IBM Quantum, AWS Braket).
- **Future-proof security**: Based on **quantum information-theoretic security**, not computational assumptions.

How BQC Works (High-Level)

The most well-known BQC scheme is based on **Measurement-Based Quantum Computing (MBQC)** and uses **entangled resource states** (e.g., cluster states):

1. **Client Prepares Qubits**
   The client prepares simple quantum states (e.g., randomly rotated single-qubit states called "qubits in disguise") and sends them to the server.

2. **Server Entangles Qubits**
   The quantum server constructs a large entangled cluster state from the received qubits.

3. **Client Sends Instructions**
   The client sends classical instructions on how to measure each qubit—but adds a random rotation that hides the actual computation.

4. **Server Performs Measurements**
   The server performs measurements and sends results back to the client.

5. **Client Decodes Result**
   The client uses the measurement results and her private angles to reconstruct the output.

Because the client's instructions are "obfuscated," and the server sees only random measurement bases, the server has **no idea what it's computing**.

Variants and Improvements

- **Universal Blind Quantum Computing (UBQC)**
  First full scheme for universal quantum computing in a blind manner. Developed by Broadbent, Fitzsimons, and Kashefi in 2009.
- **Verifiable Blind Quantum Computing (VBQC)**
  Extends BQC to allow the client to **verify** whether the server performed the computation correctly—critical for trust.
- **Classical Client BQC (Open Problem)**
  Whether a fully classical client can delegate blind computation to a quantum server securely remains a key open question. Some progress has been made using **cryptographic assumptions** and **interactive proofs**.

Real-World Use Cases

- **Secure cloud-based quantum simulation**,
- **Outsourcing quantum machine learning tasks**,
- **Quantum private query (QPQ)** and **secure database access**,
- **Blind quantum metrology** (measurement tasks without revealing the system),

- **Multi-party quantum protocols** with privacy constraints.

BQC vs Classical Homomorphic Encryption

Feature	Homomorphic Encryption	Blind Quantum Computing
Data encrypted?	Yes (mathematically)	Yes (quantum-state level)
Hardware required	Classical client/server	Quantum server; client may be semi-quantum
Function privacy	Partially (some schemes)	Fully (function and input hidden)
Security basis	Hard math problems	Quantum physics
Maturity	Higher	Experimental/research stage

Limitations and Challenges

- **Quantum-capable client required** (for most schemes),
- **Communication overhead**: Requires back-and-forth interaction,
- **Noise and decoherence**: Real systems are imperfect,
- **Scalability**: Current demonstrations are limited to small circuits.

Experimental Demonstrations

- Small-scale BQC demonstrations have been implemented using:
    - **Photonic systems** (e.g., single photons with linear optics),
    - **Trapped ions** and **superconducting circuits**,
- Proved feasibility of client-server quantum security protocols.

Security Guarantees of BQC

- **Input Blindness**: The server cannot learn the client's input.
- **Computation Blindness**: The structure of the quantum circuit is hidden.
- **Output Blindness**: Only the client can reconstruct the output.
- **Optionally Verifiable**: VBQC extensions let clients detect cheating servers.

These properties hold **unconditionally**—no assumption of computational hardness is required, only the validity of quantum mechanics.

Future of BQC

- **Practical quantum cloud security** for industries like pharma, finance, defense.

- **Integration with PQC and QKD** to build full-stack quantum-safe platforms.
- Development of **"device-independent BQC"** where even trust in the quantum device isn't needed.
- Movement toward **BQC with classical clients**, possibly via zero-knowledge protocols or trusted setups.

Conclusion: Quantum Computation Without Exposure

Blind Quantum Computing enables secure, outsourced quantum computing—making it possible to **use quantum hardware without trusting it**. In a future where quantum cloud platforms dominate, BQC will ensure that **quantum power doesn't come at the cost of privacy**.

It's a powerful vision: one where **you can compute in quantum**—without anyone ever knowing **what you computed**.

Chapter 35

Secure Multi-Party Computation with Qubits

In a world increasingly dependent on data sharing and collaboration, **Secure Multi-Party Computation (SMPC)** allows multiple parties to jointly compute a function over their private inputs—**without revealing those inputs to each other**.

Now imagine extending this privacy guarantee to the quantum realm. With **qubits**, **entanglement**, and **quantum teleportation**, new forms of SMPC emerge—potentially more powerful, private, and efficient than their classical counterparts.

Quantum Secure Multi-Party Computation (QSMPC) blends techniques from quantum cryptography, quantum communication, and distributed systems to securely compute quantum or classical functions over distributed quantum data.

What Is Secure Multi-Party Computation (Recap)?

SMPC is a cryptographic method enabling
n

n parties to compute a function
f(x1 ,x2 ,...,xn) such that:

- Each party's input ixi remains secret.
- The correct output is computed collaboratively.
- No information is leaked beyond what can be inferred from the output.

SMPC is the backbone of private voting, secure auctions, collaborative ML, and federated analytics.

Quantum SMPC: Core Idea

Quantum SMPC (QSMPC) generalizes the classical idea by allowing parties to input and manipulate **quantum states**, and compute over **quantum circuits or observables**, all while preserving quantum and classical privacy.

Two main goals:

1. **Securely evaluate a quantum circuit** across multiple parties.
2. **Prevent leakage of quantum state information** during collaborative quantum processing.

Key Building Blocks in QSMPC

1. Quantum Secret Sharing

- Splits a quantum state among parties so that **only a group together** can reconstruct it.
- Example: **(k, n)-threshold schemes**, like the **quantum generalization of Shamir's scheme**.
- Built using **entanglement and measurements**.

2. Entanglement-Based Correlation

- Parties share **entangled qubits (e.g., GHZ states)**.
- Measurements on parts of the entangled state yield correlated results.
- Used for **coordinated computation** and **non-local control**.

3. Quantum Teleportation and Gate Delegation

- Parties can **perform quantum gates remotely** by teleporting quantum states.
- Enables one party to compute on another's data **without directly accessing it**.

4. Classical Coordination + Quantum Execution

- A classical SMPC layer orchestrates protocol steps,
- The quantum layer handles **entangled resource preparation** and **circuit execution**.

Types of QSMPC Problems

- **Quantum Voting Protocols**: Secure and verifiable anonymous elections.
- **Quantum Private Set Intersection**: Find common data between parties without revealing individual sets.
- **Distributed Quantum Machine Learning**: Train a joint model without sharing raw quantum data.
- **Joint Quantum Simulation**: Run molecular or physics simulations across private quantum inputs.

Security Models:

Threat Model	Description
Honest-but-Curious	Parties follow the protocol but try to learn extra.
Malicious Adversaries	Parties may deviate to manipulate output or steal data.
Collusion	A subset of parties colludes to infer private inputs.

Quantum SMPC protocols can be designed under all these models, though the complexity increases.

Example: GHZ-Based Quantum Voting Protocol

1. Distribute a GHZ state among all voters.
2. Each voter applies a local operation encoding their vote (yes/no).
3. Collective measurement of the final state reveals the **sum** of votes, but **not individual choices**.

This offers:

- **Anonymity**,
- **Verifiability**,
- **No need for a trusted third party**.

Comparison: Classical vs Quantum SMPC

Feature	Classical SMPC	Quantum SMPC
Communication Medium	Classical channels	Classical + quantum channels
Input Types	Classical data	Classical + quantum states
Function Complexity	Boolean/circuit-based	Unitary circuits/ observables
Security Basis	Cryptographic hardness	Quantum no-cloning + entanglement
Scalability	High with optimization	Early-stage but promising

Challenges in QSMPC

- **Quantum Communication Infrastructure**: Needs reliable entanglement distribution and teleportation.
- **Noise Sensitivity**: Qubits are fragile and errors accumulate during multi-party interaction.
- **Synchronization**: Timing and coordination across quantum systems is non-trivial.
- **Protocol Complexity**: Requires expertise in both quantum information and cryptographic proofs.

Research and Progress

- **Protocols developed for specific quantum tasks**, like summation, voting, and key agreement.

- **Quantum circuit garbling** (analogous to classical garbled circuits) is being explored.
- Progress in **quantum homomorphic encryption** and **multi-party blind computation** could enhance QSMPC.

Practical Applications

- **Confidential Quantum Finance**: Joint risk models without exposing private portfolios.
- **Secure Supply Chain**: Quantum tracking of provenance without exposing proprietary data.
- **Collaborative Drug Discovery**: Shared quantum simulations between pharma competitors.
- **Quantum Auctions and Marketplaces**: Participants bid or trade without revealing strategy.

Future Outlook

QSMPC is a crucial piece in the architecture of the **quantum internet**, enabling **collaborative, distributed, and privacy-preserving quantum computation**.

As quantum networking matures—through advances in **quantum repeaters**, **entanglement routing**, and **device-independent protocols**—QSMPC will become not only feasible, but essential for secure multiparty operations in quantum cloud environments.

Conclusion: Shared Secrets in Quantum Space

Secure Multi-Party Computation with Qubits offers a vision of **collaborative trust without compromise**—where participants compute together while protecting what's theirs.

It leverages the **non-local, non-closable nature of quantum information** to achieve forms of privacy and joint action that classical systems cannot match. As we build quantum networks, QSMPC will help ensure that trust is built **into the protocol**, not assumed in the participants.

Chapter 36

Zero-Knowledge Proofs in Quantum Systems

What Is a Zero-Knowledge Proof? (Recap)

A **Zero-Knowledge Proof (ZKP)** is a cryptographic method that allows one party (the *prover*) to convince another party (the *verifier*) that a certain statement is true **without revealing why or how it's true**.

In classical terms, this means:

- The verifier gains **no additional knowledge** beyond the truth of the statement.
- The interaction reveals **nothing** about the prover's private input or secret.

ZKPs are crucial in secure identity verification, confidential transactions, and privacy-preserving protocols.

Why Extend ZKPs to Quantum Systems?

As quantum computing and communication emerge, **new questions arise**:

- How do we **verify** quantum computations without accessing the quantum data?
- Can we **prove knowledge of a quantum state** or secret without revealing it?

- How can quantum mechanics **strengthen or threaten** classical ZKPs?

These questions lead to **Quantum Zero-Knowledge Proofs (QZKPs)**—protocols that extend zero-knowledge principles into the quantum realm.

Two Main Branches of Quantum ZKPs

1. **Classical Zero-Knowledge in a Quantum World**
   - Classical ZKPs adapted to remain **secure against quantum adversaries**.
   - Ensures that post-quantum systems can't be broken by quantum computers.
2. **Genuinely Quantum ZKPs**
   - Protocols where the *statements*, *provers*, or *verifiers* are quantum.
   - May involve quantum proofs, quantum inputs, or quantum verifiers.

Classical ZKPs vs Quantum Threats

Many classical ZKPs (e.g., Schnorr protocol, zk-SNARKs) are based on **math problems vulnerable to quantum attacks** like:

- Discrete logarithms (breakable via Shor's algorithm),
- RSA and ECC-based commitments.

Thus, there's a push to build **post-quantum ZKPs** using:

- **Lattice-based cryptography**,

- **Hash-based commitments**,
- **Post-quantum signature schemes**.

These can provide **zero-knowledge security even against quantum adversaries**, preserving privacy in the post-quantum era.

Quantum Zero-Knowledge Proofs (QZKPs): Overview

Quantum ZKPs can take several forms:

- **Proving possession of a quantum state**, without revealing the state.
- **Interactive protocols using quantum messages**, where cheating verifiers can't extract information.
- **Verification of quantum computation**, where the verifier is classical but the prover is quantum.

These protocols rely on **unique quantum properties**:

- **No-cloning**: The prover can't duplicate the quantum secret.
- **Measurement disturbance**: Any probing by the verifier can be detected.
- **Entanglement**: Used to coordinate proofs or simulate consistency.

Examples of Quantum ZKP Problems

- **QMA-hard problems**: Prover convinces verifier they know a solution to a quantum problem in the QMA complexity class (quantum analogue of NP).

- **State authentication**: Prover proves knowledge of a specific quantum state (e.g., a ground state or eigenstate).
- **Blind quantum computation**: A special case where a server computes on encrypted data and proves it acted honestly.
- **Quantum money verification**: Demonstrating possession of a non-cloneable quantum state used as currency.

Interactive Quantum ZK Protocol Example (Simplified)

1. The **prover** prepares a quantum state that encodes a solution to a known hard problem.
2. The **verifier** challenges the prover to perform specific quantum operations or measurements.
3. The prover responds with results.
4. The verifier checks consistency **without learning the solution**.

If the prover cheats or doesn't know the solution, they'll be caught with high probability. If they're honest, the verifier learns nothing about the solution—only that it exists.

Key Properties of Quantum ZKPs

Property	Classical ZKPs	Quantum ZKPs
Zero-Knowledge	Hides proof content	Hides quantum state/circuit
Soundness	Cheating prover can't fool verifier	Same—using quantum info constraints

Completeness	Honest prover always succeeds	Same
Simulation	Verifier's view can be simulated	Requires quantum simulation methods

Quantum ZKPs offer **information-theoretic security** under quantum assumptions—beyond what most classical schemes can guarantee.

Applications of Quantum ZKPs

- **Quantum Authentication**: Prove ownership of a quantum key or credential.
- **Quantum Blockchain**: ZKPs ensure transaction validity without revealing identities or values.
- **Quantum-Resistant Identity**: Prove identity using quantum secrets or lattice-based credentials.
- **Secure Outsourced Quantum Computation**: Clients verify computation correctness without revealing input/output.

Challenges in QZKP Design

- **Quantum message fragility**: Noise can destroy proof information.
- **No universal ZK compiler**: Unlike classical protocols, building general-purpose QZKPs is hard.
- **Verifier's role**: Many protocols require a quantum-capable verifier—less practical currently.
- **Simulatability**: Constructing efficient quantum simulators (to prove zero-knowledge) is complex.

Recent Research Directions

- **Classical Verifier Quantum Prover (CVQP)** protocols:
    - Let a classical party verify a quantum prover.
    - Related to **Mahadev's protocol**, a landmark 2018 result enabling classical verification of quantum computation under cryptographic assumptions.
- **Quantum zk-SNARKs**:
    - Quantum analogs of succinct, non-interactive ZKPs.
    - Still theoretical but could be powerful in quantum-secure blockchains.
- **Post-Quantum Classical ZK**:
    - Development of lattice-based ZK protocols for practical digital identity and credentials.

Conclusion: Privacy Beyond Classical Boundaries

Zero-Knowledge Proofs in Quantum Systems redefine what it means to **prove something without revealing it**. By harnessing the subtleties of quantum mechanics, we unlock stronger, more nuanced privacy guarantees.

In a future filled with **quantum cloud services**, **quantum tokens**, and **quantum-secure identities**, ZKPs will be the bridge between **verifiability and privacy**—letting us say, "I know this," without ever showing what "this" is.

Section 7: Quantum Software and Programming

Chapter 37

Quantum Programming Paradigms

As quantum computing moves from theory into practical applications, a new generation of developers faces a challenge: **How do you program a quantum computer?**

Unlike classical computers, where bits and logical gates are deterministic and discrete, quantum computers operate in **complex, probabilistic, and entangled** spaces. This demands not only new hardware but also new **programming paradigms**, tools, and mental models.

Quantum programming is **not just a shift in syntax—it's a shift in thinking.**

What Is a Quantum Programming Paradigm?

A **programming paradigm** defines the structure, rules, and building blocks of how software is written. In classical computing, common paradigms include:

- **Imperative (C, Python)**,
- **Functional (Haskell)**,
- **Object-oriented (Java, C++)**,
- **Logic-based (Prolog)**.

In quantum computing, programming paradigms address how **quantum circuits**, **data encoding**, **gate operations**, and **measurements** are modeled and executed.

Key Quantum Programming Paradigms

1. Quantum Circuit Model (Gate-Based Programming)

This is the most common and hardware-agnostic paradigm. Programs are built as **quantum circuits** composed of qubit registers and unitary gates.

- **Qubits**: Act as quantum memory.
- **Gates**: Unitary transformations (e.g., H, X, CNOT, T).
- **Measurement**: Collapses quantum state to classical bits.

Languages/tools:

- Qiskit (IBM)
- Cirq (Google)
- ProjectQ
- PennyLane (Hybrid circuits)

Example:

```
qc.h(0)

qc.cx(0, 1)

qc.measure_all()
```

2. Measurement-Based Quantum Computing (MBQC)

Here, computation is performed by **preparing a large entangled state** (e.g., a cluster state), and then **measuring qubits** in specific bases.

- Computation = Measurement + Feed-forward logic.

- Used in some **blind quantum computing** and **photonic systems**.

Paradigm shift: Logic is in **measurement pattern**, not gates.

3. Quantum Functional Programming

Inspired by languages like Haskell, this paradigm treats quantum programs as **pure functions** with no side effects.

- Emphasizes **immutability**, **recursion**, and **algebraic types**.
- Supports **reasoning about quantum effects** through category theory or lambda calculus.

Examples:

- Quipper (developed in Haskell)
- Q# (Microsoft) incorporates some functional features.

4. Hybrid Quantum-Classical Programming

Focuses on integrating quantum components with classical ML models or control loops. Widely used in **variational quantum algorithms (VQAs)**.

- Classical host code controls quantum subroutines.
- Quantum parts act as accelerators or feature mappers.

Languages/tools:

- PennyLane (Xanadu)
- TensorFlow Quantum
- Qiskit Runtime

5. Quantum Logic Programming

Still largely theoretical, this paradigm uses **quantum predicates** and **logical inference** on quantum states.

- Analogous to Prolog, but for quantum queries and constraints.
- Potential use in quantum databases or reasoning systems.

Programming Models vs Execution Models

Programming Paradigm	Execution Model	Use Case
Gate-based (circuit)	Gate-model quantum hardware	General quantum computation
MBQC	One-way photonic computers	Blind computation, security
Functional	Simulated or compiled	Reversible logic, theoretical work
Hybrid (classical host)	Variational quantum hardware	VQE, QAOA, ML

Common Features of Quantum Programming

- **No Loops in Quantum Hardware**: Loops must be unrolled or managed classically.
- **No Cloning**: You cannot copy a qubit—data duplication must be handled carefully.

- **Probabilistic Outputs**: Measurement doesn't always yield the same result.
- **State Vector or Density Matrix**: Backend may simulate wave function or mixed states.

Quantum Programming Languages: Overview

Language	Paradigm	Backend Support	Key Features
Qiskit	Gate-based	IBM Quantum	Open-source, modular
Cirq	Gate-based	Google (Sycamore)	Optimized for NISQ
Q#	Structured/ Functional	Azure Quantum	Strong type safety, simulator-friendly
PennyLane	Hybrid (ML)	Any backend	Autograd, VQAs, quantum ML
Quipper	Functional	Haskell-based	Circuit description + algebra

Debugging Quantum Programs

Debugging is **non-trivial** due to:

- No peeking at intermediate states,
- Probabilistic outcomes,
- Complex amplitude spaces.

Tools used:

- Simulators (e.g., Aer, Qsim),
- Circuit visualizers,
- State vector inspection (only in simulation),
- Tomography (in real hardware).

Challenges for Developers

- **Mindset shift**: Determinism → probability.
- **Limited debugging tools**.
- **Noise and decoherence** in real devices.
- **Limited quantum RAM** (qRAM is not practical yet).
- **Low-level abstraction**: Most languages are still close to the hardware layer.

Toward Higher-Level Quantum Abstractions

Efforts are underway to build:

- **Quantum compilers** (e.g., t|ket⟩, Quilc),
- **Quantum intermediate representations** (QIR by Microsoft),
- **DSLs for quantum algorithms** (domain-specific languages),
- **Cross-platform APIs** (OpenQASM 3.0, Qiskit Metal for chip-level design).

Conclusion: Programming in Quantum Dimensions

Quantum programming paradigms are evolving rapidly, pushing the boundaries of computation and abstraction. Whether you're constructing circuits, orchestrating hybrid models, or reasoning about entanglement, your programming toolkit must adapt to the **rules of quantum physics**.

It's not just about writing code—it's about **thinking quantum.**

Chapter 38

Qiskit, Cirq, and PennyLane: Tools of the Trade

The growth of quantum computing has been accompanied by the rise of powerful software frameworks that help developers **write**, **simulate**, and **execute** quantum algorithms. Among them, **Qiskit**, **Cirq**, and **PennyLane** have emerged as leading tools, each with unique strengths tailored to specific hardware backends and programming paradigms.

Whether you're designing circuits, building quantum machine learning models, or running hybrid algorithms, choosing the right tool is essential. Let's explore how these platforms stack up—**in architecture, design philosophy, and real-world use cases.**

1. Qiskit (IBM)

Overview:
Qiskit is IBM's open-source quantum computing SDK, built to design and execute gate-based quantum circuits on simulators or real IBM Q hardware.

Strengths:

- Rich suite of **modular components** (Terra, Aer, Ignis, Aqua).
- Native support for **IBM Quantum processors**.
- Tight integration with **OpenQASM**, a quantum assembly language.

- Built-in tools for **visualization, simulation, and transpilation**.

Typical Use Cases:

- Quantum algorithm prototyping (e.g., Grover, Shor).
- VQE and QAOA with Qiskit Nature or Qiskit Optimization.
- Quantum education (via Qiskit Textbook).

Developer Experience:

```
from qiskit import QuantumCircuit

qc = QuantumCircuit(2)

qc.h(0)

qc.cx(0, 1)

qc.measure_all()
```

2. Cirq (Google)

Overview:
Cirq is Google's quantum framework optimized for **NISQ devices**, particularly its Sycamore hardware. It's lightweight and **hardware-near**, built for maximum control.

Strengths:

- Designed for **fine-grained pulse-level control**.
- Clear support for **NISQ experiments** (noisy intermediate-scale quantum).
- Integration with **TensorFlow Quantum** and **quantum supremacy experiments**.

- Native focus on **circuit fidelity and optimization**.

Typical Use Cases:

- Experimental quantum algorithms on Sycamore.
- Benchmarking and error mitigation.
- Circuit simulation on local backends.

Developer Experience:

```
import cirq

qubits = [cirq.GridQubit(0, i) for i in range(2)]

circuit = cirq.Circuit(cirq.H(qubits[0]), cirq.CNOT(qubits[0],
qubits[1]))
```

3. PennyLane (Xanadu)

Overview:
PennyLane is a **quantum machine learning (QML)** and **hybrid quantum-classical computing** platform that integrates seamlessly with PyTorch, TensorFlow, and JAX.

Strengths:

- First framework to support **automatic differentiation** across quantum circuits.
- Perfect for **VQE**, **QAOA**, and **quantum neural networks**.
- Plug-and-play with many backends (Qiskit, Cirq, Braket, Strawberry Fields).
- Built for **hybrid workflows** with classical training loops.

Typical Use Cases:

- Training quantum classifiers and generative models.
- Implementing variational circuits with classical optimizers.
- Exploring near-term quantum ML.

Developer Experience:

```
import pennylane as qml

dev = qml.device("default.qubit", wires=2)

@qml.qnode(dev)

def circuit(params):

    qml.RY(params[0], wires=0)

    qml.CNOT(wires=[0, 1])

    return qml.expval(qml.PauliZ(1))
```

Comparative Summary:

Feature	Qiskit	Cirq	PennyLane
Maintainer	IBM	Google	Xanadu
Main Focus	General-purpose quantum	NISQ experimentation	Hybrid quantum-classical ML
Backend Compatibility	IBMQ, simulators	Google Sycamore, local	Many (via plugins)
Strengths	Circuit construction, transpilation	Hardware-level fidelity	ML, autodiff, hybrid loops
Target Audience	Researchers, educators	Experimentalists, physicists	ML practitioners, QML devs

Classical Integration	Moderate (Pythonic)	Low	High (TF, PyTorch, JAX)
Community & Docs	Strong (Qiskit Textbook)	Strong for Google users	Growing rapidly

Choosing the Right Tool

- Use **Qiskit** if you want a full-stack platform to build quantum circuits, simulate them, and run on **real IBM hardware**. Excellent for education, research, and prototyping.
- Choose **Cirq** if you're close to hardware and care about **NISQ constraints**, circuit timing, and running on **Google's quantum chips**.
- Go with **PennyLane** if your goal is to explore **quantum machine learning**, **differentiable programming**, or **hybrid quantum-classical systems** with tools like PyTorch or TensorFlow.

Looking Ahead

As the ecosystem matures, these frameworks are becoming increasingly **interoperable**, with initiatives like **OpenQASM 3.0** and **QIR** aiming to create **portable quantum programs**.

Eventually, developers may be able to write once and run anywhere—on a **diverse quantum cloud** powered by Qiskit, Cirq, PennyLane, and beyond.

Chapter 39

Writing Efficient Quantum Code

In classical computing, efficient code saves memory, time, and power. In quantum computing, it does something even more vital: it **enables programs to run at all**.

With quantum processors still limited in qubit count, coherence time, and gate fidelity, writing **efficient quantum code** isn't just good practice—it's a survival skill. Each operation, qubit, and moment of runtime matters.

This chapter explores key principles, patterns, and tools for designing quantum circuits and hybrid algorithms that are as **lightweight, optimized, and noise-resilient** as possible.

Why Efficiency Is Crucial in Quantum Code

- **Qubits are scarce**: Most devices support fewer than 100 usable qubits.
- **Decoherence limits runtime**: A quantum program must finish before qubit states degrade.
- **Gate errors accumulate**: The more gates you apply, the higher the noise.
- **Compilers and mappers may add depth**: Inefficient code often gets worse after transpilation.

Efficient quantum code helps you:

- Minimize circuit depth (runtime),

- Reduce qubit usage (space),
- Limit gate count (error probability),
- Increase chance of correct measurement.

Key Strategies for Efficient Quantum Programming

1. Minimize Circuit Depth

Depth = number of sequential gate layers. A shallow circuit finishes faster and is more robust.

Tips:

- Parallelize independent operations.
- Use **commuting gates** strategically.
- Combine multiple gates into equivalent unitaries (e.g., gate fusion).
- Avoid unnecessary basis changes (e.g., H or T gates).

2. Reduce Total Qubit Count

Fewer qubits = less chance of decoherence + more portability.

Tips:

- Use ancilla qubits only when absolutely needed.
- Exploit **qubit reuse** via uncomputation and measurement.
- Consider **mid-circuit measurement and reset**.

3. Use Native Gate Sets

Each quantum device has its own **native gate set**. Using non-native gates can introduce translation errors and extra operations.

Tips:

- Match your circuit to the target backend's gate set (e.g., CNOT + RX/RZ for IBM).
- Use device-aware transpilation.

4. Optimize Entanglement Usage

Entanglement is powerful but error-prone.

Tips:

- Limit long-range entanglement across physical qubits.
- Use swap networks carefully (for nearest-neighbor constraints).
- Decompose multi-controlled gates (e.g., Toffoli) efficiently.

Common Patterns of Inefficient Quantum Code

Anti-Pattern	Problem	Efficient Alternative
Deep nested CNOT chains	Accumulates error	Optimize for topology
Overuse of ancilla qubits	Consumes limited qubit resources	Uncompute + reuse

Global entanglement	Physically unrealistic or costly	Local entanglement + SWAPs
Redundant measurements	Adds noise + consumes bandwidth	Use only when needed

Use Transpilation and Compilation Wisely

Most platforms provide **transpilers** to optimize circuits post-design.

Example with Qiskit:

```
from qiskit import transpile

optimized_circuit = transpile(qc, basis_gates=['u3', 'cx'], optimization_level=3)
```

Levels range from 0 (no change) to 3 (aggressive optimizations like gate fusion, layout selection, routing, and resynthesis).

Efficient Hybrid Algorithms (e.g., VQE, QAOA)

For hybrid quantum-classical loops, efficiency extends beyond the circuit:

- **Use gradient-aware optimizers** (like Adam or SPSA).
- **Batch quantum evaluations** when possible.
- **Reduce circuit width and depth per ansatz layer**.
- **Warm-start** classical parameters from prior runs.
- **Cache classical computations** (e.g., matrix elements).

Use Parameterized Gates and Templates

Avoid hardcoding parameters when running VQAs or iterative solvers.

Instead of:

circuit.ry(0.45, 0)

Do:
circuit.ry(theta[0], 0)

This allows reuse of compiled circuits with new parameters, reducing compilation time and backend calls.

Simulation-Time Optimizations

- Use **state vector simulator** only when necessary—slow and memory-intensive.
- Prefer **density matrix simulators** for noise modeling.
- Use **unitary simulators** for pure circuit logic debugging.

Debugging for Efficiency

- **Visualize circuits** with `draw()` or `circuit_drawer`.
- **Estimate circuit cost** using:
    - Qiskit's `circuit.depth()`, `size()`, `width()`
    - Cirq's `circuit.moment_by_moment_stats()`

Example:

print("Circuit depth:", qc.depth())

print("Number of gates:", qc.size())

Benchmark and Profile

- Measure real backend fidelity before execution.
- Run **calibration-aware transpilation** using backend-specific error maps.
- Profile hybrid loops to locate classical bottlenecks.

Toolkits That Help

Toolkit	Feature	Ideal For
Qiskit Terra	Transpilation, metrics, layout	IBM backends, optimization
Cirq optimizers	Circuit merging, qubit mapping	Google hardware simulations
PennyLane	Parameter shift, autograd	Hybrid QML workflows
**t	ket⟩**	Cross-platform transpiler

Conclusion: Think Like a Quantum Compiler

Writing efficient quantum code demands not only understanding the algorithm but also the **hardware**, **noise model**, and **execution path**. It's a balance between logic, layout, and physics.

Good quantum programmers are **part physicist, part compiler**, and **part algorithm designer**. With the right habits and tools, you can turn limited quantum resources into practical quantum advantage—one optimized circuit at a time.

Chapter 40

Simulating Quantum Circuits

Before we can run a quantum program on a real device, we often simulate it to check its logic, visualize outputs, and benchmark performance. **Quantum circuit simulation** provides a powerful way to model quantum behavior on classical computers—offering developers and researchers a safe, efficient, and noise-free sandbox.

Though simulations are limited in scale, they're indispensable for algorithm development, education, and verification.

Why Simulate Quantum Circuits?

- **Design and test algorithms** without needing quantum hardware access.
- **Visualize quantum states and operations** step by step.
- **Debug and optimize circuits** before deploying to noisy hardware.
- **Model ideal vs noisy behavior** for benchmarking.
- **Compare different algorithms** with controlled variables.

Types of Quantum Simulators

Simulator Type	Description	Use Case
Statevector	Computes full quantum state vector	Ideal, small circuits (≤20 qubits)
Unitary	Simulates the full unitary matrix of a circuit	Gate logic inspection
Density Matrix	Models decoherence and mixed states	Noisy simulation
Stabilizer	Efficient for Clifford-only circuits	Large-scale stabilizer algorithms
Tensor Network	Contract tensors to scale beyond 30–40 qubits (approx.)	Quantum chemistry, condensed matter
Noisy Hardware Emulators	Simulate hardware-specific gate errors, noise, delays	Realistic execution preview

State vector Simulation: The Most Common Approach

A **state vector simulator** calculates the quantum state at each step of a circuit.

Qiskit Example:

```
from qiskit import Aer, execute
from qiskit.visualization import plot_state_city

backend = Aer.get_backend('statevector_simulator')
result = execute(circuit, backend).result()
```

```
state vector = result.get_statevector()
```

Useful for:

- Inspecting amplitudes and phase relationships,
- Understanding quantum interference,
- Verifying expected output superpositions.

⚠️ Limit: **Scales exponentially**. 20 qubits = 1 million amplitudes.

Noisy Simulation: Adding Real-World Behavior

Use simulators with **noise models** to evaluate algorithm performance under realistic conditions.

Qiskit Example:

```
from qiskit.providers.aer.noise import NoiseModel

noise_model = NoiseModel.from_backend(backend)

result = execute(circuit, Aer.get_backend('qasm_simulator'),

             noise_model=noise_model).result()
```

Helps test:

- Error rates,
- Decoherence effects,
- Mitigation strategies.

Measurement-Based Simulation (QASM)

Simulates final measurements only (as real hardware would do).

- Returns classical outcomes.
- Useful for circuits involving randomness, such as Grover's algorithm or QAOA.

Example:

```
circuit.measure_all()

backend = Aer.get_backend('qasm_simulator')

result = execute(circuit, backend, shots=1024).result()

counts = result.get_counts()
```

Simulators by Framework

Framework	Simulators Available
Qiskit	`statevector_simulator`, `qasm_simulator`, `unitary_simulator`, `density_matrix_simulator`
Cirq	`Simulator`, `DensityMatrixSimulator`, `CliffordSimulator`
PennyLane	Plugin-based simulators (default.qubit, lightning.qubit, etc.)
Forest SDK	`WavefunctionSimulator`, `ReferenceQVM` (Rigetti)

Tips for Effective Quantum Simulation

- **Keep circuits shallow**: Especially for full-state simulators.
- **Track resource use**: Memory use scales exponentially with qubit count.
- **Use noise models for realism**: Simulate IBM, Google, or custom devices.
- **Benchmark variants**: Try ideal vs noisy circuits for performance gaps.
- **Visualize outputs**: Use state vector plots, histograms, and Bloch spheres.

Advanced Techniques

- **Snapshot Instructions** (Qiskit): Extract mid-circuit data for analysis.
- **Parallel Simulation**: Use high-performance computing (e.g., IBM's Qiskit Aer with MPI).
- **Tensor Network Simulation**: Tools like ITensor, Cotengra, or QFlex simulate 50+ qubits under structure constraints.

Limitations of Classical Simulation

- Cannot simulate arbitrary quantum circuits at large scale due to **exponential memory requirements**.
- Infeasible to model realistic noise in circuits >20 qubits accurately.
- Simulators are **idealized** unless explicitly modeled with noise.

Use Cases in Practice

Use Case	Simulation Method
Algorithm design (VQE)	Hybrid + statevector
ML model prototyping	PennyLane `default.qubit`
Grover's search testing	QASM simulation
Noise testing	Noisy QASM / density matrix
Teaching and demos	Statevector + visualizers

Conclusion: Virtual Labs for Quantum Computing

Simulating quantum circuits gives developers a **laboratory without hardware**—a crucial sandbox to iterate, test, and learn. Whether you're designing from scratch or verifying results before running on a quantum device, simulators help bridge the gap between theory and execution.

While real quantum advantage requires physical machines, simulation remains our most powerful tool for **building the future now.**

Chapter 41

Quantum Compilation and Optimization

Quantum programming doesn't stop at writing code. Before any quantum circuit can run on real hardware, it must go through a process of **compilation and optimization**—a series of steps that translate, rearrange, and tailor your circuit to the quirks of a specific quantum device.

Just as a C++ program is compiled into machine instructions optimized for a CPU architecture, quantum circuits must be **mapped, minimized, and matched** to a quantum processor's physical constraints.

What Is Quantum Compilation?

Quantum compilation is the process of transforming a high-level quantum circuit into a low-level form that can be executed efficiently on a specific quantum backend.

It involves:

- **Gate decomposition** (into native gates),
- **Topology-aware mapping** (qubit layout matching),
- **Gate scheduling and routing**,
- **Error mitigation strategies**,
- **Circuit simplification and optimization**.

Compilation Pipeline Overview

1. **High-Level Quantum Code**
   - e.g., A circuit written in Qiskit, Cirq, or PennyLane.
2. **Intermediate Representation (IR)**
   - e.g., OpenQASM, Quil, QIR.
3. **Gate Decomposition**
   - Convert high-level gates (like Toffoli) into primitive gates (e.g., CX, RZ, RX).
4. **Qubit Mapping**
   - Match logical qubits to physical ones on the hardware.
5. **Routing and Swap Insertion**
   - Add SWAP gates when qubits aren't physically adjacent.
6. **Gate Optimization**
   - Cancel redundant gates, fuse adjacent rotations, etc.
7. **Backend-Specific Translation**
   - Convert into device-native instructions.
8. **Execution**

Why Optimization Is Essential

Quantum devices are constrained by:

- **Limited connectivity** between qubits,
- **Gate errors** and **decoherence**,
- **Restricted gate sets**,
- **Short coherence times**.

A poorly optimized circuit:

- Takes longer to run,
- Has more errors,
- May exceed device limits,
- Produces lower-fidelity results.

Key Optimization Techniques

Technique	Purpose
Gate Fusion	Combine sequential rotations into one
Gate Cancellation	Eliminate inverse or redundant gates
Commutation Analysis	Reorder gates to enable optimization
SWAP Minimization	Reduce overhead of moving qubits

Pulse-Level Optimization	Fine-tune for hardware-specific pulse timing
Measurement Grouping	Merge measurements to minimize shots

Hardware-Aware Compilation

Each quantum processor has a **native gate set** and **qubit topology**:

- IBM's QPUs: CX, U3; hex lattice topology
- Google's Sycamore: CZ, single-qubit gates; grid
- Rigetti's Aspen: Native CZ and RX/RZ; square lattice

The compiler adjusts:

- **Which gates to use** (e.g., convert H → RX + RZ),
- **Which qubits to place where**,
- **When to insert SWAPs** for routing.

Tools and Compilers

Tool	Description
Qiskit Terra	Includes transpiler and optimization levels

Cirq Optimizers	Includes circuit transformers and routers
**t	ket⟩ (Cambridge Quantum)**
Quilc (Rigetti)	Compiler for Forest SDK, supports Quil
Xanadu Compiler (PennyLane)	Optimizes variational circuits

Qiskit Transpiler Example

```
from qiskit import transpile
optimized_circuit = transpile(circuit,
backend=ibmq_backend, optimization_level=3)
```

Level 0: Minimal change (preserve structure)

Level 1: Basic gate mapping

Level 2: Moderate optimization (gate merging, layout)

Level 3: Aggressive optimization

Metrics to Watch

Metric	Meaning
Gate Count	Total number of gates
Circuit Depth	Max number of sequential layers

Two-Qubit Gates	Count of CNOTs/CZs—most error-prone
SWAPs Inserted	Due to qubit topology constraints
Execution Fidelity	Estimate of final circuit quality

Quantum IRs (Intermediate Representations)

IRs are abstract languages that enable cross-platform compilation:

- **OpenQASM 3.0**: IBM's intermediate format with classical logic
- **Quil**: Rigetti's quantum instruction language
- **QIR**: Microsoft's LLVM-based quantum IR
- **Tket IR**: Intermediate circuit format for t|ket⟩

These IRs are used to:

- Enable cross-compiler optimization,
- Simulate and benchmark,
- Port code across quantum platforms.

Post-Compilation Verification

After compilation, it's critical to:

- **Check equivalence** between original and compiled circuit,
- **Simulate** the final circuit to test fidelity,
- **Visualize** mapping (e.g., qubit layout and connectivity),
- **Compare metrics** before and after optimization.

Challenges in Quantum Compilation

- **Balancing depth and gate fidelity**,
- **Limited hardware knowledge** (e.g., pulse shapes, decoherence profiles),
- **Compiler biases** (some optimizations hurt specific use cases),
- **Dynamic circuits**: Harder to optimize in advance due to feedback/control.

Future of Quantum Compilation

- **AI-based optimization**: Using reinforcement learning to improve mapping.
- **Hardware-in-the-loop compilers**: Use live data to tune routing.
- **Pulse-aware compilation**: Co-design software with pulse-level physics.
- **Cross-platform abstraction**: Write once, compile anywhere.

Conclusion: The Compiler Is Your Co-Pilot

Quantum compilation and optimization transform raw circuits into **high-performance instructions** that match the real-world limits of today's quantum computers. With thoughtful compilation, even noisy, limited machines can deliver valuable results.

A good compiler doesn't just translate—it **amplifies your algorithm's effectiveness**, bringing quantum advantage one gate closer.

Chapter 42

Debugging and Testing Quantum Software

Quantum software development is unlike classical development in many ways—but perhaps most striking is the challenge of **debugging**. In classical systems, debugging tools let you inspect memory, track variable values, and step through code. But in quantum systems, **the act of observation changes the state**—or destroys it altogether.

This makes **testing and debugging quantum code a unique discipline**, requiring new tools, techniques, and mindsets.

Why Is Debugging Quantum Software So Hard?

- **Measurement collapse**: Reading a qubit's state collapses its superposition.
- **No cloning theorem**: You can't copy quantum data for comparison or logging.
- **Probabilistic outcomes**: Repeated runs may yield different results.
- **Opaque quantum state**: Intermediate states can't be examined directly on hardware.
- **Noise and decoherence**: Even "correct" code may fail due to real-world imperfections.

Despite these challenges, debugging is not only possible—it's critical to building reliable quantum programs.

Core Testing Goals

1. **Correctness** – Does the circuit implement the intended logic?
2. **Fidelity** – How close is the final state to the ideal?
3. **Robustness** – Does the circuit behave well under noise?
4. **Efficiency** – Are resources (qubits, gates, depth) used optimally?

Techniques for Debugging Quantum Programs

✅ 1. Use Simulators Before Hardware

Simulators allow full access to:

- **State vectors**
- **Unitary matrices**
- **Intermediate measurement results**

You can simulate an ideal quantum environment and inspect the **exact amplitudes**—something impossible on real hardware.

```
from qiskit import Aer

result = execute(circuit,
Aer.get_backend("statevector_simulator")).result()

print(result.get_statevector())
```

✅ 2. Visualize Circuits

Visual tools help catch structural issues:

- **Gate order**
- **Entanglement structure**
- **Unintended measurements or resets**

Qiskit:

```
circuit.draw("mpl")  # Matplotlib style
```

PennyLane and Cirq also offer similar visualization functions.

✅ 3. Test with Known Inputs and Outputs

Design **unit tests** that prepare known states and check if expected measurements or state vectors match.

Example:

- Apply Hadamard to |0⟩ ⇒ Expect 50/50 between |0⟩ and |1⟩.
- Prepare a Bell state ⇒ Check entanglement correlation.

✅ 4. Use Assertion Circuits

In some frameworks, you can build **custom assertion gates** or measurement-check subcircuits.

Qiskit Ignis:

- Check for parity, Hamming weight, or fidelity thresholds.
- Use statistical testing over multiple shots.

✅ 5. Test Intermediate States (Sim Only)

Split your circuit into parts:

- Simulate and inspect each part individually.
- Compare expected and simulated vectors.

You can also use **snapshot** functionality (Qiskit) to inspect mid-circuit states in a simulator.

✅ 6. Use Tomography (Real Hardware)

On real quantum hardware, **quantum state tomography** and **process tomography** help reconstruct what happened:

- **State tomography**: Rebuilds a qubit's density matrix from many measurements.
- **Process tomography**: Reveals how a quantum gate or circuit behaved.

Useful for small subsystems due to exponential scaling.

✅ 7. Classical Shadowing

Use **classical shadow techniques** to estimate a quantum state or observable efficiently from random measurements. This reduces the number of shots needed and increases testing feasibility on larger systems.

Test Types in Quantum Software

Test Type	Purpose	Example
Unit Test	Check individual gate or circuit correctness	Hadamard → 50/50 test
Integration Test	Validate subcircuits interacting as expected	QAOA or VQE layer behavior
Fidelity Test	Measure how close final state is to expected	Compare statevectors
Regression Test	Ensure fixes don't break prior performance	Validate metrics post-refactor
Noise Tolerance Test	Evaluate behavior under errors	Run on noisy simulator

Tools for Debugging and Testing

Tool	Feature
Qiskit Ignis	Error mitigation + test circuits
Qiskit Aer	Noise simulation + snapshots
Cirq Simulators	Moment-by-moment inspection
PennyLane Lightning	Fast gradient evaluation
Pytest	For classical-quantum test integration
Statevector Comparison	For ideal test checking

Best Practices for Reliable Quantum Code

- Write modular code with testable subcircuits.
- Run simulations with ideal and noisy models.
- Always validate against expected results when possible.
- Track qubit layout and mapping when targeting real hardware.

- Use transpilation to match device constraints **after** logic testing.

Common Debugging Mistakes

Mistake	Consequence	Fix
Ignoring noise in simulation	Unrealistic performance	Use noise models
Forgetting to measure	Output is undefined	Explicitly add `measure_all()`
Over-measuring in hardware	Collapses useful states too early	Delay measurement until needed
Assuming simulator = hardware	Code may break in reality	Validate on real backend or emulated noise

Conclusion: Quantum Debugging Is an Art

You can't inspect quantum memory, but you can build confidence in your software with **careful simulation, structured testing, and probabilistic reasoning**. Debugging quantum programs requires creativity, curiosity, and a new mental model—**more like detective work than step-by-step tracing**.

With the right tools and habits, you can catch bugs early, build trust in your quantum code, and write software that works both in theory and on hardware.

Section 8: Applications in Science and Industry

Chapter 43

Quantum Chemistry and Molecular Modeling

Quantum chemistry aims to understand the behavior of atoms and molecules by solving the Schrödinger equation for multi-electron systems. However, solving this equation exactly for large molecules is **exponentially hard**—even for the best classical supercomputers.

This is where **quantum computing** enters the stage.

Quantum computers, by nature, **simulate quantum systems**. With qubits and quantum gates, they can represent molecular orbitals and electronic interactions far more naturally and efficiently than classical methods. **Quantum chemistry** is poised to become one of the first domains where quantum computers offer a **real computational advantage**.

The Classical Bottleneck in Chemistry

- **Electron correlation**: Describing how electrons avoid each other in atoms/molecules is computationally intense.
- **Scaling**: Methods like full configuration interaction (FCI) scale as O

  (O(N!).
- **Trade-offs**: High-accuracy methods (e.g., coupled-cluster) are limited to small molecules, while faster methods sacrifice precision.

Quantum Chemistry on Quantum Computers

Quantum computers provide a new framework:

- **Encode electrons into qubits** using mappings (Jordan-Wigner, Bravyi-Kitaev).
- **Construct a Hamiltonian** for the molecule (second quantization).
- **Apply quantum algorithms** to find the ground state energy or simulate dynamics.

Main Quantum Algorithms for Chemistry

✅ Variational Quantum Eigensolver (VQE)

- Hybrid quantum-classical algorithm.
- Quantum circuit prepares a parameterized trial wave function (ansatz).
- Classical optimizer updates parameters to minimize the expected energy.

Ideal for NISQ devices due to short, parameterized circuits.

✅ Quantum Phase Estimation (QPE)

- Uses quantum coherence and interference to find eigenvalues (e.g., energy levels).
- High accuracy but requires deep circuits—suited for fault-tolerant future.

Qubit Mappings for Molecular Systems

Mapping	Description	Qubit Cost
Jordan-Wigner	Straightforward, qubits = orbitals	Moderate
Bravyi-Kitaev	More efficient parity mapping	Lower depth
Parity Mapping	Hybrid of the above	Balanced

These map molecular orbitals and electron operators into qubit operations.

Workflow for Quantum Molecular Simulation

1. **Molecular structure input**: Choose molecule (e.g., H_2, LiH, FeMoCo).
2. **Classical pre-processing**: Compute one- and two-electron integrals using tools like PySCF or Psi4.
3. **Hamiltonian generation**: Translate into second-quantized form, then map to qubits.
4. **Choose ansatz** (UCCSD, hardware-efficient, etc.).

5. **Run VQE/QPE** on simulator or quantum hardware.

6. **Interpret energy**: Compare with known ground state, excited states, or chemical properties.

Quantum Chemistry Platforms

Platform	Highlights
Qiskit Nature	Integrates with PySCF, supports VQE, QPE
OpenFermion	From Google; chemistry + Cirq integration
PennyLane + PySCF	Hybrid ML + chemistry simulation
Orquestra (Zapata)	Chemistry pipelines, cloud-ready

These platforms help automate Hamiltonian conversion, ansatz construction, and optimization.

Example: Hydrogen Molecule with VQE (Qiskit)

```
from qiskit_nature.drivers import PySCFDriver

from qiskit_nature.problems.second_quantization.electronic
import ElectronicStructureProblem

driver = PySCFDriver(atom='H .0 .0 .0; H .0 .0 0.735',
basis='sto3g')

problem = ElectronicStructureProblem(driver)
```

- Maps the molecule to a 4-qubit Hamiltonian.
- Runs VQE with a UCC ansatz and a COBYLA optimizer.

Use Cases and Impact Areas

- **Catalyst design**: Simulate reaction intermediates and transition states.
- **Drug discovery**: Calculate binding energies, molecular interactions.
- **Materials science**: Study strongly correlated materials like high-Tc superconductors.
- **Photosynthesis and nitrogen fixation**: Quantum models for complex biological systems.

Challenges and Limitations

- **Qubit count**: Accurate modeling of medium-size molecules needs 100–1000 qubits.
- **Error rates**: Chemistry simulations are sensitive to gate and readout noise.
- **Ansatz design**: Must balance expressiveness and circuit depth.
- **Hamiltonian scaling**: 2-electron integrals explode in number as system grows.

Near-Term Success Stories

- **H_2, LiH, BeH_2**: Simulated with high accuracy on real quantum hardware.

- **Zapata + BMW**: Quantum chemistry for battery design.
- **IBM + ExxonMobil**: Modeling hydrogen and lithium-based reactions.

These early cases demonstrate the **feasibility and promise** of quantum chemistry.

The Road Ahead

- **Fault-tolerant QPE** will enable accurate simulation of transition metals, catalysts, and proteins.
- **New ansatzes** and adaptive algorithms (e.g., ADAPT-VQE) will improve efficiency.
- **Tighter integration with classical tools** (ML, HPC) will accelerate the design loop.
- **Cross-disciplinary collaboration** between chemists, physicists, and quantum engineers will be essential.

Conclusion: Molecules Meet Qubits

Quantum chemistry is one of the clearest applications of quantum computing with transformative implications. From **drug discovery** to **clean energy**, quantum simulation of molecules may unlock breakthroughs that were simply impossible before.

With careful engineering and continuous innovation, **quantum computers will soon become indispensable microscopes**, allowing us to see—and shape—chemistry at its deepest level.

Chapter 44

Quantum Optimization for Logistics and Finance

Optimization problems are everywhere: routing trucks, allocating assets, scheduling flights, pricing derivatives, managing risk. These problems grow rapidly in complexity with scale—often becoming **NP-hard**, meaning no efficient classical solution is known.

Quantum computing offers **new algorithms and approaches** that can potentially solve such problems **faster or more effectively**, using the power of **superposition**, **entanglement**, and **quantum parallelism**.

In this chapter, we explore how quantum optimization can enhance **logistics and finance**—two industries where even a slight improvement in optimization translates into massive cost savings and competitive advantage.

Why Quantum for Optimization?

Classical optimization techniques—like branch-and-bound, simulated annealing, and mixed-integer programming—struggle with:

- **Combinatorial explosion**,
- **Multiple local minima**,
- **Real-time re-optimization** in dynamic environments.

Quantum computers, especially near-term NISQ devices, offer new strategies such as:

- **Quantum Approximate Optimization Algorithm (QAOA)**,
- **Quantum annealing**,
- **Hybrid quantum-classical solvers**.

Quantum Optimization Techniques

✅ 1. Quantum Approximate Optimization Algorithm (QAOA)

- Maps a classical optimization problem into a quantum cost function (Hamiltonian).
- Alternates between applying problem-specific and mixing operators.
- Parameters are optimized classically to maximize solution probability.

Great for:

- Max-Cut,
- Graph coloring,
- Vehicle routing,
- Portfolio balancing.

✅ 2. Variational Quantum Algorithms (VQAs)

- More flexible framework than QAOA.
- Uses parametrized circuits to minimize cost functions (e.g., risk, delivery time).

✅ 3. Quantum Annealing

- Inspired by physical annealing, quantum annealers like D-Wave **explore solution space via tunneling**.
- Ideal for combinatorial problems expressible as Ising models.

✅ 4. Quantum-Inspired Algorithms

- Run on classical hardware but use **quantum-inspired heuristics** like tensor networks and low-rank approximations.

Applications in Logistics

Problem	Quantum Approach	Outcome
Vehicle Routing	QAOA / annealing	Fewer miles, faster deliveries
Supply Chain Optimization	QAOA + ML	Resilient, cost-optimized networks
Crew Scheduling	VQA / hybrid	Fewer conflicts, better utilization
Warehouse Picking	QAOA + reinforcement learning	Optimal item routing

Quantum optimization can consider **multiple constraints** (fuel, time, labor, carbon) simultaneously—something classical heuristics often struggle to balance well.

Applications in Finance

Problem	Quantum Approach	Outcome
Portfolio Optimization	QAOA / VQE	Better return-risk trade-offs
Risk Management	Hybrid VQE	Faster scenario simulations
Options Pricing	Quantum Monte Carlo	Fewer samples, accurate estimates
Fraud Detection	Quantum ML + optimization	Faster pattern recognition

Quantum optimization helps:

- Handle **non-convex, nonlinear objective functions**,
- Incorporate **correlated assets and dynamic market changes**,
- Explore **large solution spaces** faster than brute-force methods.

Industry Use Cases and Pilots

- **Volkswagen + D-Wave**: Traffic flow optimization in Beijing.
- **Goldman Sachs + QC Ware**: Risk modeling and Monte Carlo acceleration.
- **HSBC + IBM Quantum**: Portfolio analysis and hedging strategies.
- **CERN + Cambridge Quantum**: Scheduling high-volume data workflows.

These collaborations reflect growing confidence in hybrid quantum optimization.

Strengths and Challenges

Strengths	Challenges
Solves problems with many variables	Quantum hardware still limited
Natural fit for combinatorics	Noise affects accuracy
Hybrid models improve practicality	Classical solvers still dominate
Quantum speedups are possible	Problem mapping can be complex

Quantum advantage in optimization isn't universal yet—but **for structured problems** with the right encoding, it's emerging.

Quantum Optimization Workflow

1. **Define the problem** (cost, constraints, objective).
2. **Map it to a quantum Hamiltonian** (e.g., QUBO form).
3. **Choose a quantum algorithm** (QAOA, VQE, annealing).
4. **Build parameterized circuits**.
5. **Run classical optimization loop** to minimize or maximize.
6. **Measure results** and interpret bitstrings as solutions.

Tools like **Qiskit Optimization**, **D-Wave Ocean SDK**, and **PennyLane QAOA modules** simplify this process.

Tools for Quantum Optimization

Platform	Highlights
Qiskit Optimization	Convex/Ising problem modeling + QAOA
D-Wave Ocean	Quantum annealing tools for logistics
PennyLane + PyTorch	Hybrid VQA optimization pipelines
Amazon Braket	Access to multiple quantum hardware + hybrid solvers

Best Practices for Practitioners

- Start with **small problem instances** to validate encoding.
- Benchmark against **classical solvers**.
- Use **noise-aware simulators** to refine hybrid circuits.
- Combine quantum results with **ML predictions** for adaptive logistics or trading.
- Evaluate solution **quality and stability**, not just raw speed.

Looking Ahead: Quantum-Driven Optimization Ecosystems

- **Quantum scheduling engines** will optimize smart factories and global shipping networks.
- **Real-time quantum trading assistants** may rebalance portfolios in milliseconds.
- **Autonomous logistics platforms** will adapt on the fly using quantum inference.

As quantum hardware scales and matures, industries that rely on heavy optimization will **experience step-change improvements in performance, accuracy, and adaptability**.

Conclusion: Better Decisions, Faster

Quantum optimization represents a paradigm shift for decision-making. In logistics and finance, where complexity, uncertainty, and scale challenge traditional methods, quantum solutions offer a new way to **explore possibilities, reduce costs, and respond faster** to change.

By blending quantum algorithms with classical infrastructure, we move closer to a world of **precision operations at quantum speed**.

Chapter 45

Drug Discovery and Materials Science

Modern drug design and materials innovation depend heavily on understanding how atoms and molecules interact. This involves solving complex quantum mechanical problems to simulate molecular behavior, protein-ligand binding, and electronic structures—tasks that often push classical computing to its limits.

Quantum computing offers a transformative leap forward, enabling more accurate and scalable simulation of the very quantum systems that define biology and materials. With quantum tools, scientists can go beyond approximation, entering a future where **molecular innovation is accelerated, personalized, and far more precise**.

Why Quantum Computing Matters Here

- **Biological molecules and materials obey quantum rules.**
- Current tools rely on **approximations** (e.g., DFT, MD), which can break down in complex or correlated systems.
- Quantum computers, in principle, can simulate **quantum behavior natively**—opening doors to new discoveries and optimizations.

Key Challenges in Drug and Material Design

Area	Classical Bottleneck
Drug Discovery	Binding affinity prediction, docking

Protein Folding	Massive conformational space
Materials Science	Modeling electronic, magnetic, thermal properties
Battery Chemistry	Transition metal modeling, phase stability

These problems often involve:

- Large Hilbert spaces,
- Strong electron correlations,
- Long simulation times.

How Quantum Computing Helps

✅ 1. Molecular Simulation

- Simulate small molecules (H_2, LiH, FeMoCo) to compute ground-state energies, reaction pathways, and orbital interactions.
- Replace approximate methods with **quantum exactness** (via VQE or QPE).

✅ 2. Drug-Target Binding Prediction

- Quantum machine learning can enhance binding affinity scoring.
- Simulate binding site environments with quantum-enhanced models.

✅ 3. Protein Folding

- Use quantum annealing or hybrid optimization to navigate the protein energy landscape.
- Search for minimal free-energy structures.

✅ **4. Materials Property Modeling**

- Determine band structures, conductivity, magnetism with quantum-accurate models.
- Simulate **strongly correlated systems**, where classical tools fail.

Algorithms Used

Algorithm	Application Area
Variational Quantum Eigensolver (VQE)	Molecular ground state energy
Quantum Phase Estimation (QPE)	High-precision simulation
Quantum Annealing	Protein folding, molecular docking
Quantum Kernel Methods	Drug activity classification
Quantum Neural Networks (QNNs)	Predicting drug efficacy, side effects

Real-World Initiatives and Collaborations

- **Roche + Cambridge Quantum**: Quantum models for molecular simulation.
- **Pfizer + IBM**: VQE and quantum chemistry for drug R&D.

- **Zapata + BASF**: Quantum workflows for materials discovery.
- **QC Ware**: Quantum-enhanced property prediction and simulation tools.

These partnerships reflect a **shared goal: reduce the time and cost of R&D cycles** using quantum acceleration.

Platforms and Tools

Platform	Specialization
Qiskit Nature	Quantum chemistry and materials modeling
OpenFermion	Electronic structure calculations
PennyLane	Hybrid quantum ML for property prediction
Orquestra	Drug discovery pipelines
Menten AI	Quantum protein engineering tools

These platforms bridge **quantum hardware** with **domain-specific software workflows** in chemistry, biology, and physics.

Example Workflow: Quantum Drug Candidate Screening

1. Select a target protein structure.

2. Generate molecular candidates (ligands).
3. Compute approximate quantum properties (energy, orbital interaction).
4. Use VQE to refine candidate energetics.
5. Rank and select top binders using quantum ML classifiers.

Such pipelines could eventually **replace traditional docking and scoring workflows**, particularly for novel or complex targets.

Impact in Materials Science

- **Battery design**: Simulate cathode/anode behavior, ionic diffusion.
- **Semiconductors**: Predict defect behavior and conductivity under stress.
- **Superconductors**: Model electron pairing in high-Tc materials.
- **Catalysts**: Understand surface reactions and reaction rates.

Quantum simulation will help **optimize materials at the atomic scale**, reducing trial-and-error in labs.

Current Limitations

- Qubits and coherence time still limit size of molecules that can be modeled (typically ~20 orbitals).
- Noise impacts accuracy for long-depth algorithms like QPE.

- Preprocessing (e.g., Hartree-Fock) still requires powerful classical systems.
- End-to-end pipelines are early-stage and experimental.

Future Directions

- **Fault-tolerant quantum chemistry** with 1,000+ logical qubits will unlock full protein-ligand models.
- **Adaptive ansatzes (ADAPT-VQE)** will reduce circuit size while increasing accuracy.
- **Quantum ML + molecular dynamics** will create new hybrid simulation paradigms.
- **Automated design of molecules** using generative quantum models (e.g., QGANs).

Conclusion: Chemistry Meets the Quantum Age

Quantum computing is poised to **redefine how we discover drugs and design materials**. What once required years of lab work and simulation may be reduced to **hours of quantum computation**, guided by algorithms that can explore, simulate, and optimize molecules at a fidelity never seen before.

As hardware improves and tools mature, quantum-enhanced R&D will move from promise to production—ushering in a **golden age of innovation in chemistry and materials science**.

Chapter 46

Climate Modeling and Energy Simulations

Climate change is one of the most urgent challenges facing humanity. To understand and respond effectively, scientists rely on **climate models**—vast simulations of the Earth's atmosphere, oceans, biosphere, and human impact. Likewise, transitioning to renewable energy and optimizing energy systems depend on accurately simulating **complex, multi-scale processes** in physics, chemistry, and economics.

But today's models are often **limited by classical computing**. They must approximate critical processes like cloud formation, fluid dynamics, and material behaviors at scale—leading to trade-offs between **resolution, accuracy, and computational cost**.

Quantum computing offers a potentially revolutionary alternative: **physics-native simulation**, exponential parallelism, and new hybrid algorithms to model Earth's climate and energy systems **with greater realism, speed, and efficiency**.

Why Classical Models Struggle

Domain	Bottlenecks
Climate Prediction	Resolution limits, cloud microphysics
Ocean Currents	Non-linear Navier-Stokes equations

Renewable Grid Modeling	Combinatorial complexity, stochastic inputs
Fusion & Fission	Particle interactions, quantum plasma

These simulations require:

- Massive parallelism,
- Multi-scale resolution (nanometers to kilometers),
- Handling of chaotic, coupled systems.

How Quantum Computing Can Help

✅ 1. Solving Partial Differential Equations (PDEs)

Climate models rely on PDEs for dynamics (e.g., fluid flow, temperature transport). Quantum algorithms like **Quantum Linear Solvers** or **Hamiltonian simulation** can solve sparse linear systems exponentially faster under ideal conditions.

✅ 2. Turbulence and Quantum Fluid Dynamics

Quantum simulations can model turbulence and complex fluid interactions more naturally via **quantum lattice gas algorithms** and **quantum walk-based fluid models**.

✅ 3. Material and Energy Simulations

- Simulate **photovoltaic materials**, hydrogen storage systems, or superconductors using quantum chemistry.
- Study catalysts for **CO_2 capture**, battery materials, and **fusion plasma behaviors**.

✅ 4. Optimization of Energy Networks

Use **Quantum Approximate Optimization Algorithm (QAOA)** and hybrid models for:

- Smart grid scheduling,
- Renewable dispatch,
- Load forecasting,
- Energy trading strategy.

Quantum Use Cases in Climate and Energy

Application Area	Quantum Approach
Cloud microphysics	Quantum-enhanced PDE solvers
CO_2 capture materials	Quantum chemistry (VQE/QPE)
Weather prediction	Quantum ML + data assimilation
Wind farm optimization	Quantum optimization + hybrid ML
Grid balancing	QAOA + classical forecasting
Fusion reactor modeling	Quantum simulation of plasma and transport phenomena

Platforms and Partnerships

- **Pasqal + EDF**: Energy grid and climate-aware network optimization.
- **NASA + QCI**: Climate model acceleration via quantum PDE solvers.
- **TotalEnergies + QC Ware**: Hydrogen and catalyst quantum simulations.
- **IBM + Electric Power Research Institute (EPRI)**: Hybrid quantum models for grid resilience.

These efforts reflect a growing global commitment to use quantum computing for sustainability.

Hybrid Quantum-Classical Climate Pipelines

1. **Collect observational data** (satellite, ocean, sensor networks).
2. **Run classical pre-processing and coarse modeling.**
3. **Deploy quantum solvers** for key sub-models (e.g., fluid flow, material reactions).
4. **Optimize** energy or mitigation strategies using hybrid loops.
5. **Feed results into AI-enhanced policy or adaptation tools.**

Quantum Algorithms Relevant to Climate

Algorithm	Use Case

Quantum Linear Systems Algorithm (QLSA)	Solving sparse systems in PDEs
Variational Quantum Simulation	Model time-dependent processes
Quantum Walks	Particle and fluid simulation
Quantum Boltzmann Machines	Weather pattern recognition
Quantum Neural Networks	Spatiotemporal forecasting

Challenges and Future Opportunities

Challenge	Outlook
Large-scale climate requires thousands of qubits	Fault-tolerant computing in 5–10 years
Noise sensitivity in iterative solvers	Near-term hybrid error mitigation
Integration with HPC systems	Emerging quantum-HPC workflows
Model interpretability	Quantum ML for causal discovery

Vision for a Quantum-Powered Planet

- **Real-time flood prediction and wildfire modeling** using quantum-enhanced simulations.
- **Global carbon cycle models** with full-chemistry interactions.
- **Next-generation batteries and carbon-neutral fuels** designed via quantum chemistry.
- **Smart, adaptive energy grids** balancing millions of variables daily.

Quantum computing offers not just performance gains—it enables **fundamentally new ways to model and manage the complexity of Earth itself**.

Conclusion: Quantum Climate Intelligence

As humanity navigates the dual challenges of climate change and sustainable energy, quantum computing may become one of our most powerful tools. By simulating nature with unprecedented precision, quantum computers can help us make smarter decisions, design cleaner systems, and build a more resilient future.

From clouds to catalysts, from grid nodes to global policies—**quantum technologies will soon be woven into the fabric of Earth system science**.

Chapter 47

Quantum Advantage in Supply Chain Problems

In today's global economy, supply chains are the arteries of commerce—connecting suppliers, manufacturers, warehouses, retailers, and customers across vast networks. These systems are incredibly complex and sensitive to delays, disruptions, and cost fluctuations.

Optimizing supply chains involves **multi-variable decision-making under uncertainty**, spanning procurement, production, inventory, transportation, and logistics. Classical methods (e.g., MILP, heuristics) can only go so far—especially when problems grow combinatorially large.

Quantum computing, with its ability to explore massive solution spaces and model complex constraints, offers a compelling path forward. From route planning to demand forecasting, quantum algorithms may provide a **computational advantage** in key supply chain areas where classical approaches struggle.

Why Supply Chain Optimization Is So Hard

- **NP-hard problems**: Many tasks (e.g., vehicle routing, facility layout) are computationally intractable as scale increases.
- **Dynamic inputs**: Real-time demand, weather, delays, and market conditions constantly change.
- **Multimodal constraints**: Must optimize for cost, time, emissions, and service levels simultaneously.

- **Global interdependence**: Local changes ripple through interconnected systems.

These realities demand tools capable of **adaptive, large-scale optimization under uncertainty**—an area where quantum computing is poised to help.

Where Quantum Advantage May Emerge

Problem Area	Quantum Approach	Benefit
Inventory optimization	QAOA, VQE	Balances holding and shortage costs
Vehicle routing (VRP)	QAOA, annealing	Optimized last-mile delivery
Production scheduling	Hybrid quantum solvers	Reduces idle time and delays
Supplier selection	Quantum ML + optimization	Adaptive sourcing recommendations
Warehouse picking	Quantum walk, annealing	Efficient order fulfillment paths
Demand forecasting	Quantum-enhanced ML	More accurate short-term predictions

Quantum Algorithms in Action

✅ 1. Quantum Approximate Optimization Algorithm (QAOA)

- Used to solve constraint-based combinatorial problems by encoding them into a cost Hamiltonian.
- Alternates between "problem" and "mixer" Hamiltonians to explore solution space.

Use Case: Optimize multi-drop delivery routes while minimizing time, fuel, and carbon emissions.

✅ 2. Quantum Annealing (e.g., D-Wave)

- Maps optimization problems onto an energy landscape and finds low-energy (optimal) solutions through quantum tunneling.
- Particularly useful for QUBO (Quadratic Unconstrained Binary Optimization) models.

Use Case: Shift scheduling with labor and legal constraints.

✅ 3. Variational Quantum Algorithms

- Combines parameterized quantum circuits with classical optimizers to minimize cost functions.

Use Case: Balancing inventory costs with service level constraints across global warehouses.

Real-World Pilots and Initiatives

- **Volkswagen + D-Wave**: Traffic flow optimization for taxis in Beijing.
- **Mercedes-Benz**: Exploring battery supply chain modeling using quantum optimization.

- **Zapata + BMW**: Multi-objective logistics optimization under uncertainty.
- **Accenture + Biogen**: Quantum modeling of supply and demand dynamics for pharmaceutical distribution.

These pilots show quantum readiness in high-impact supply chain sectors.

Quantum Supply Chain Workflow Example

1. **Formulate problem** as QUBO or cost Hamiltonian.
2. **Map variables** (e.g., routes, inventory, shifts) to qubits or parameters.
3. **Choose solver** (QAOA, VQE, annealing).
4. **Run hybrid loop**: classical optimizer updates quantum circuit.
5. **Extract solution**: interpret bitstring or parameter set.
6. **Validate** against KPIs (e.g., cost, time, service level).

Hybrid Models for Practical Deployment

Because current quantum devices are still limited, **hybrid models** are essential. These combine:

- Classical pre-processing (data cleaning, clustering),
- Quantum core optimization,
- Classical post-processing (scoring, policy simulation).

Example: Use ML to forecast demand clusters, then QAOA to optimize supplier dispatch across regions.

Platform and Tools

Tool/Platform	Capability
Qiskit Optimization	Formulates and solves QUBO problems
D-Wave Ocean SDK	Models real-world logistics as Ising models
PennyLane + TensorFlow	Hybrid ML + optimization pipelines
Amazon Braket	Multi-hardware access + simulators

Potential Benefits at Scale

Metric	Improvement Potential
Total logistics cost	↓ 5–20% through route & inventory gains
Order cycle time	↓ 15–30% via better scheduling
On-time delivery	↑ 10–25% with adaptive routing
Carbon footprint	↓ with optimized transportation

Disruption resilience	↑ through faster re-optimization

Challenges and Road Ahead

- **Hardware limitations**: Large problems still require more qubits and better error correction.
- **Encoding complexity**: Mapping real-world constraints into quantum-friendly formats is hard.
- **Data integration**: Connecting ERP systems with quantum pipelines is non-trivial.
- **Interpretability**: Translating quantum outputs into supply chain KPIs needs standardization.

Despite these, **quantum supply chain optimization is already being prototyped at enterprise levels**—with promising early results.

Conclusion: Quantum-Enabled Agility

In a world where disruptions are frequent and agility is key, quantum optimization offers **a competitive edge** in managing supply chain complexity. While full-scale deployment will take time, even near-term quantum advantage in niche problems can drive millions in savings and improvements.

Quantum computing is not replacing your ERP—**it's making it smarter, faster, and more adaptive**.

Chapter 48

Quantum Advantage in Supply Chain Problems

Financial Risk Modeling with Quantum Algorithms

Financial risk is all about uncertainty. Banks, hedge funds, insurers, and regulators need to estimate the likelihood and impact of adverse events—market crashes, interest rate shifts, defaults, or cyber threats. These assessments depend on **complex probabilistic models**, large data sets, and computationally intense simulations.

Classical methods like Monte Carlo simulations and copula models are effective but scale poorly, especially when modeling **tail risks**, **nonlinear dependencies**, and **multivariate stress scenarios**.

Quantum algorithms—particularly quantum Monte Carlo, amplitude estimation, and hybrid optimization—are showing **real promise in enhancing both the speed and depth** of financial risk modeling.

Why Quantum Risk Modeling?

Classical Limitation	Quantum Solution
Slow convergence of Monte Carlo	Quantum Amplitude Estimation (QAE)
High-dimensional integrals	Quantum-enhanced integration
Portfolio VaR across scenarios	QAOA or VQE optimization

Difficult correlations (copulas)	Quantum machine learning
Simultaneous shock propagation	Quantum parallelism over distributions

Quantum computing improves risk analysis by offering **faster convergence**, **better modeling of uncertainty**, and **efficient handling of high-dimensional spaces**.

Key Quantum Algorithms for Risk Modeling

✅ 1. Quantum Amplitude Estimation (QAE)

- Core algorithm to estimate expected values (means, probabilities) quadratically faster than classical Monte Carlo.
- Uses **quantum interference and Grover iterations** to accelerate convergence.

Use Case: Faster estimation of Value at Risk (VaR) and Conditional Value at Risk (CVaR).

✅ 2. Quantum Monte Carlo

- Implements stochastic sampling using quantum circuits.
- Enhances simulation of uncertain financial processes (e.g., asset paths, credit events).

✅ 3. Quantum Variational Algorithms

- VQE/QAOA can solve optimization problems with risk-adjusted objective functions (e.g., minimize downside risk subject to returns).

✅ **4. Quantum Machine Learning**

- Quantum-enhanced kernel methods and quantum neural networks can detect patterns in historical risk data (e.g., for early warning signals).

Use Cases in Financial Risk

Application Area	Quantum Algorithm	Outcome
VaR & CVaR estimation	QAE, QMC	Faster, more accurate tail estimates
Stress testing	Hybrid VQA	Simulation under multi-variable shocks
Credit risk modeling	QML for default probability	Better clustering of credit profiles
Liquidity risk	Quantum graph search	Rapid identification of bottlenecks
Operational risk	Quantum causal discovery	Uncover systemic interdependencies

Quantum Value at Risk (VaR) Estimation

VaR is the threshold loss not exceeded with a given confidence over a time horizon.

Classical Monte Carlo:

- Requires millions of simulations to estimate rare-event tail risks.

Quantum Advantage:

- QAE reduces complexity from O(1/ϵ2) to O(1/ε) is the error tolerance.

This enables:

- Faster risk reporting,
- Higher-fidelity tail analysis,
- Real-time revaluation of positions.

Hybrid Quantum-Classical Risk Workflows

1. **Collect historical data**: Prices, volatilities, exposures.
2. **Build classical preprocessing models**: Factor models, scenario generators.
3. **Encode risk measures into quantum circuits**: Payoff functions, cost Hamiltonians.
4. **Run quantum algorithm**: QAE, QAOA, or hybrid model.
5. **Analyze outputs**: Distributions, percentiles, stress responses.
6. **Feed into decision engines**: Capital allocation, hedging, compliance.

Example: Quantum Risk Modeling with Qiskit Finance

```
from qiskit_finance.applications.estimation import
CVaRMeasurement
```

Define uncertainty model, payoff function, and use QAE to estimate CVaR

IBM's Qiskit Finance includes modules for:

- Log-normal asset modeling,
- Option pricing,
- CVaR estimation,
- Portfolio optimization under risk constraints.

Industry Collaborations

- **Goldman Sachs + QC Ware**: Fast CVaR estimation via QAE.
- **HSBC + IBM Quantum**: Risk model acceleration for capital stress tests.
- **JPMorgan + Qiskit**: Quantum simulation of option pricing and volatility paths.
- **Fitch + Zapata**: Credit risk modeling using quantum-inspired techniques.

These projects aim to reduce **risk quantification time** from hours to minutes, or enable **more detailed simulations** with the same resources.

Challenges in Quantum Risk Modeling

Challenge	Status/Progress

Noise in QAE	Mitigated via error-aware iterations
Encoding complex payoffs	Needs circuit design heuristics
Data privacy/compliance	Still emerging legal framework
Real-time integration	Dependent on hardware and APIs

Risk modeling needs highly reliable estimates—making **error mitigation** and **hardware stability** essential for adoption.

Benefits of Quantum Risk Modeling

Benefit	Impact
Faster risk estimation	Near real-time portfolio rebalancing
Greater tail fidelity	Better stress resilience
Enhanced scenario coverage	Higher-quality regulatory reports
Capital efficiency	Optimized reserve and liquidity buffers

As financial systems become more automated, the ability to model complex risks quickly becomes a **strategic advantage**.

The Road Ahead

- **Fault-tolerant QAE** will power real-time portfolio risk engines.

- **Quantum-native risk metrics** could replace or augment classical indicators.
- **Quantum ML + graph theory** may expose new contagion channels in markets.
- **AI + quantum fusion** will create intelligent agents that understand and hedge risk autonomously.

Conclusion: A New Frontier for Financial Safety Nets

Quantum algorithms won't eliminate financial risk—but they can make it **more visible, understandable, and manageable**. By transforming how we model uncertainty, quantum computing will help institutions respond faster, allocate capital smarter, and build systems that are resilient by design.

In a world where milliseconds can cost millions, **quantum speed in risk modeling may soon become a necessity—not a luxury**.

Section 9: Challenges and Future Directions

Chapter 49

Scalability and Hardware Bottlenecks

Quantum computing holds immense promise—but to deliver real-world impact at scale, it must overcome some of the most daunting engineering and physics challenges in modern science. As we attempt to move from 50 to 5,000 and eventually millions of qubits, the issue of **scalability** becomes central.

Despite exciting breakthroughs in algorithms and prototypes, **quantum hardware remains the greatest bottleneck** to widespread adoption. Understanding these limitations—and the strategies being developed to overcome them—is critical for technologists, business leaders, and researchers alike.

What Does "Scalability" Mean in Quantum Computing?

Scalability refers to the ability to:

- **Increase the number of high-quality qubits** without significant performance loss.
- **Maintain low error rates** across larger, deeper circuits.
- **Support complex quantum programs reliably** through hardware, control, and software integration.

A scalable system must offer **coherence, connectivity, stability, and fault tolerance** at scale—not just raw qubit counts.

Current Hardware Limitations

Bottleneck	Description
Qubit Quality	Limited by decoherence, gate fidelity, and noise
Qubit Connectivity	Sparse or fixed-topology links increase routing overhead
Error Rates	Two-qubit gates remain noisy (1–5% on most devices)
Readout Fidelity	Measurement errors distort results
Cooling Requirements	Superconducting systems require dilution fridges near 15 mK
Control Electronics	Scalable, synchronized control is difficult at cryogenic levels
Fabrication Yield	Physical variation leads to faulty or unstable qubits

These limitations hinder execution of deep or large quantum circuits and reduce algorithmic fidelity.

Coherence and Decoherence: The Time Constraint

- **T_1 (Relaxation Time)**: Time before a qubit loses its quantum state.
- **T_2 (Dephasing Time)**: Time before relative phases between states become randomized.

Typical coherence times:

- **Superconducting qubits**: ~50–200 μs
- **Trapped ions**: ~1–10 s (much longer but slower gates)

To execute meaningful algorithms, all computations must complete **within this coherence window**, or risk being overwritten by noise.

Qubit Technologies and Scalability Trade-Offs

Technology	Pros	Challenges
Superconducting	Fast gates, strong industrial momentum	Cryogenic, dense wiring complexity
Trapped Ions	Long coherence, high fidelity	Slow gate speed, limited parallelism
Photonic Qubits	Room temperature, ideal for networking	Hard to generate and entangle
Spin Qubits	CMOS-compatible, scalable in theory	Coherence still under research
Topological Qubits	Fault tolerance by design (e.g., Majorana)	Still unproven in practice

Each approach has unique trade-offs that affect **how quickly and how far they can scale**.

Error Correction: A Key to Scale

Quantum error correction is **not optional**—it's essential for building reliable quantum computers.

- Requires **logical qubits** made from hundreds or thousands of physical qubits.
- Popular codes: **Surface Code**, **Color Code**, **Bacon-Shor**.
- Thresholds: Below ~1% error per gate, fault-tolerant computation becomes feasible.

Example:

- To run Shor's algorithm on a 2,048-bit number, we may need **millions of physical qubits** with today's error rates.

Control and Wiring Bottlenecks

- Quantum processors need high-frequency microwave pulses to manipulate qubits.
- Each qubit must be **individually controlled**, leading to a wiring explosion.
- Cryogenic bottlenecks: Amplifiers, DACs, and filters must work at ultra-low temperatures.

Emerging solutions:

- **Cryo-CMOS**: Control chips that work at cryogenic temperatures.
- **Multiplexing**: Reduce wiring by sharing control lines.

Fabrication and Yield Challenges

Like classical chips, quantum chips suffer from:

- **Material defects**,

- **Inconsistent qubit frequencies**,
- **Cross-talk between qubits**.

Fabricating large arrays of high-fidelity qubits at high yield remains a major bottleneck.

Roadmaps and Scaling Strategies

Company/ Project	Scaling Approach
IBM	Modular superconducting chips (Heron, Kookaburra)
Google	Surface code + Cryo-stack integration
IonQ	Trapped ion scaling via optical routing
PsiQuantum	Fault-tolerant photonic architecture
D-Wave	Quantum annealing graph expansion

Most roadmaps involve:

- **Modularity**: Linking small chips into larger systems.
- **Error-corrected logical layers**: Shielding users from raw hardware instability.

The Power vs. Qubit Quality Trade-Off

Many early systems scale qubit count quickly—but with low fidelity.

Quality > Quantity: A 1,000-qubit noisy device may be less useful than a 100-qubit stable one. Quantum advantage depends on **both**.

Emerging Solutions and Future Trends

- **Quantum interconnects**: Photonic links between chips for modular scaling.
- **AI-enhanced calibration**: Adaptive tuning of qubit parameters and gates.
- **Topological qubits**: Intrinsic error resilience, if experimentally realized.
- **Software-hardware co-design**: Optimize algorithms for real device quirks.

Benchmarking Scalability

Metric	Goal (near-term)
Gate fidelity	>99.9% for two-qubit gates
Coherence time	>1 ms on large qubit counts
Qubit connectivity	All-to-all or high effective degree
Qubit count (logical)	100+ logical qubits post correction
Control scalability	Support 10,000+ individually addressable qubits

Conclusion: The Long Road to Quantum Supremacy

Achieving quantum advantage at scale is not just about algorithms—it's about building **machines that work at scale, under noise, and in real time**.

The road to scalable quantum computing will be paved with breakthroughs in:

- Materials science,
- Cryogenics,
- Microwave engineering,
- Chip design,
- Error correction.

We're not there yet—but each step forward brings us closer to a future where **scalable quantum computers reshape industry, science, and society**.

Chapter 50

Standardization in Quantum Technologies

As quantum computing transitions from research to real-world applications, it faces a familiar challenge seen in the early days of classical computing: fragmentation. Dozens of hardware platforms, programming languages, noise models, and performance metrics exist—many incompatible with one another.

Without **standardization**, the ecosystem risks duplication, vendor lock-in, and slow adoption. With it, quantum technologies can **interoperate**, developers can build with confidence, and businesses can scale with clarity.

Standardization isn't just a technical convenience—it's the foundation for **trust, scalability, and commercial viability** in quantum computing.

Why Standardization Matters

Area	Why It Matters
Software development	Avoids code rewrites across platforms
Benchmarking	Enables fair performance comparisons
Hardware integration	Simplifies interfacing with quantum devices

Security and compliance	Essential for finance, healthcare, defense
Industrial adoption	Reduces risk and improves ROI

Standardization fosters **ecosystem growth** by reducing complexity and aligning global innovation efforts.

Core Domains Requiring Standards

1. **Quantum Programming Languages & SDKs**
   - Common IR (intermediate representation) for cross-platform execution.
   - Examples: OpenQASM, QIR (Quantum Intermediate Representation), Quil.
2. **Hardware Abstraction & APIs**
   - Define how software interfaces with various qubit technologies.
   - Goal: Run the same code on superconducting, ion-trap, or photonic platforms.
3. **Noise Models & Error Reporting**
   - Standard formats for describing gate errors, decoherence rates, and crosstalk.
   - Enables accurate simulation, benchmarking, and hardware comparison.
4. **Benchmarking & Performance Metrics**
   - Cross-platform definitions of:

        - Qubit fidelity
        - Quantum volume
        - Circuit depth
        - Logical qubit lifetime

5. **Data Interchange & Serialization**
    - Formats for quantum data exchange (e.g., circuit structure, results, parameters).
    - Compatible with classical and quantum hybrid workflows.

Notable Standardization Initiatives

Initiative	Scope
OpenQASM (IBM)	Defines quantum programs in low-level IR
QIR Alliance (Microsoft)	LLVM-based IR for quantum hardware
QuTiP & OpenPulse	Pulse-level control standardization
IEEE P7130	Establishing quantum terminology standards
QED-C (Quantum Economic Development Consortium)	Interoperability frameworks

ISO/IEC JTC 1/SC 27	Quantum cryptography and security standards

These efforts represent a **global push toward interoperability and clarity**.

Standardizing Quantum Software Stacks

Most quantum software stacks follow this layered structure:

[User Application]

↓

[Quantum Programming Language] ← e.g., Qiskit, Cirq, PennyLane

↓

[Intermediate Representation (IR)] ← OpenQASM, QIR, Quil

↓

[Compiler & Optimizer]

↓

[Hardware API/Driver] ← e.g., Qiskit Runtime, Amazon Braket

↓

[Quantum Hardware] ← Superconducting, Trapped Ion, etc.

A standard IR ensures **portability**, while standard APIs and result formats enable **interchangeability**.

Metrics That Need Standardization

Metric	Purpose
Quantum Volume	General hardware capability
Algorithmic Qubits	Qubits usable after error correction
Gate Fidelity	Quality of individual gates
T_1 / T_2 times	Coherence and dephasing durations
Circuit Runtime	Execution latency for end-to-end jobs
Compilation Overhead	Code efficiency on hardware

Without agreed-upon definitions and test protocols, vendors and researchers cannot **compare performance fairly**.

Benefits of Standardization

✅ **Cross-Platform Compatibility**
Write once, run anywhere—reducing vendor lock-in and cost.

✅ **Reliable Benchmarks**
Allow governments and enterprises to assess tech maturity clearly.

✅ **Accelerated Development**
Tooling, optimization, and automation become easier with uniform formats.

✅ **Trust & Certification**
Critical for adoption in regulated industries like finance and defense.

✅ **Enables Quantum Cloud & Networking**
Standards allow distributed systems to cooperate, e.g., in hybrid cloud setups or quantum internet protocols.

Challenges in Achieving Standardization

- **Rapid innovation**: Hardware and algorithms evolve too fast for standards to keep up.
- **Vendor competition**: Companies may resist standardizing features that provide differentiation.
- **Hardware heterogeneity**: Superconducting, photonic, trapped ion systems all behave differently.
- **Lack of real-world usage data**: Makes it difficult to define what "typical" performance is.

Despite this, momentum is building to **define flexible, extensible frameworks** that grow with the field.

Outlook: What Comes Next

- **Global standards bodies** (like IEEE and ISO) will begin certifying compliance.
- **Open-source compilers and runtimes** will converge toward shared formats.
- **Standard testing suites** (e.g., for QAOA or VQE) will benchmark devices objectively.

- **Inter-cloud compatibility**: Run jobs across IBM, AWS, and Google seamlessly.
- **Quantum-safe cryptography** will rely on agreed protocols for key exchange and encryption.

Conclusion: Building a Quantum Future on Common Ground

The promise of quantum computing can't be realized in isolation. As hardware advances and applications multiply, **standardization will be the glue** that connects innovators, users, industries, and governments.

It will turn today's fragmented landscape into a unified, interoperable ecosystem—paving the way for widespread, responsible, and impactful quantum deployment.

Standards aren't about limiting innovation—they're about unlocking it for everyone.

Chapter 51

Quantum Internet and Communication Networks

Just as the classical internet transformed how we store, share, and secure information, a **quantum internet** could revolutionize how we **transmit quantum information** across space and time.

Instead of bits, a quantum internet will transmit **qubits**—units of quantum information that enable unbreakable encryption, distributed quantum computing, and entangled sensor networks. This new communication paradigm is not just faster or more powerful; it's **fundamentally different**, built on principles like **entanglement**, **superposition**, and **no-cloning**.

A functional quantum internet will be a **backbone for scalable quantum computing, ultra-secure communication, and novel physics experiments**—reshaping both technology and geopolitics.

What Is a Quantum Internet?

A **quantum internet** is a network that allows **quantum devices to exchange qubits** over long distances, preserving their quantum properties (coherence, entanglement).

Key goals:

- Share **entangled states** between remote parties.
- Distribute **quantum cryptographic keys** securely.
- Connect quantum computers to build **distributed quantum systems**.

- Enable **quantum-enhanced sensing and metrology**.

Why Classical Networks Can't Handle Qubits

Limitation	Why It Matters
No-cloning theorem	Qubits can't be copied to recover lost data
Quantum decoherence	Qubits degrade quickly over long distances
Loss in optical fibers	Limits how far photons can travel unassisted

Quantum communication requires **new infrastructure** and protocols to maintain fragile quantum states during transmission.

Core Components of Quantum Networks

1. **Quantum Nodes**
    - Devices (e.g., quantum computers or memory banks) that create, store, or process qubits.
2. **Quantum Channels**
    - Optical fibers or free-space links transmitting photons carrying qubits.
3. **Quantum Repeaters**

- Devices that **extend communication distance** via entanglement swapping and purification—necessary due to signal loss and decoherence.

4. **Entanglement Distribution Protocols**
    - Protocols to create and share entangled qubits across nodes, e.g., using Bell pairs.

5. **Classical Side Channel**
    - Required for coordination, measurement outcomes, and error correction in quantum teleportation.

Key Technologies

Technology	Role in Quantum Networking
Single-photon sources	Send discrete quantum states
Quantum memories	Temporarily store qubits at nodes
Entangled photon pairs	Enable teleportation and QKD
Quantum repeaters	Overcome distance limitations
Superconducting circuits & ion traps	Local processing nodes

Quantum Communication Protocols

✅ **1. Quantum Key Distribution (QKD)**

- Secure key exchange leveraging quantum physics (e.g., BB84, E91 protocols).
- Detects any eavesdropping inherently.
- Deployed today in financial and defense applications.

✅ **2. Quantum Teleportation**

- Transmits a qubit's state using shared entanglement and classical communication.
- Fundamental to **quantum repeaters** and distributed quantum computing.

✅ **3. Entanglement Swapping**

- Chains multiple entangled links to create **long-distance entanglement**.
- Enables scalable quantum networks beyond fiber attenuation limits.

Current Status of the Quantum Internet

Project/Organization	Achievement
QUANT-NET (Europe)	Fiber-based QKD between multiple cities
DARPA Quantum Network	First quantum network prototype (USA)
China's Micius satellite	Satellite-based QKD and teleportation

Japan's NICT QKD network	Tokyo-area metro QKD infrastructure
QuTech (Netherlands)	Entanglement-based networking between cities

Some efforts already connect **quantum nodes over hundreds of kilometers**, combining ground and satellite infrastructure.

Use Cases and Applications

Application Area	Description
Quantum Cryptography	Unbreakable key exchange (QKD)
Distributed Quantum Computing	Share quantum tasks across nodes
Entangled Sensor Networks	Coordinate ultra-sensitive devices
Secure Voting & Communications	Proven confidentiality and integrity
Quantum Cloud Services	Connect users to remote quantum hardware

Challenges to Overcome

Challenge	Status
Photon loss over distance	Limited to ~100 km without repeaters
Quantum memory fidelity	Still limited in duration and scale

Scalability of repeaters	Requires complex entanglement management
Global standardization	Interoperability between quantum hardware
Integration with classical networks	Synchronization and hybrid control needed

Despite these, pilot projects prove **feasibility**, and rapid R&D is narrowing the gap toward a scalable, usable network.

Toward a Quantum Internet Stack

Like the OSI model for classical internet, a **quantum internet stack** is emerging:

[Application Layer] – Quantum apps, QKD protocols

[Transport Layer] – Entanglement routing, reliability

[Network Layer] – Node addressing, switching

[Link Layer] – Entanglement generation, fidelity control

[Physical Layer] – Photons, optical fiber, satellite links

Projects like **Quantum Internet Alliance (EU)** and **QIR (Quantum Internet Router)** are working on standardized protocols and stack architectures.

The Future of Global Quantum Connectivity

- **Hybrid classical-quantum networks** will offer seamless security and computation services.

- **Quantum satellites and intercontinental links** will enable encrypted messaging and distributed sensing on a planetary scale.
- **Quantum internet access** may one day be as ubiquitous as Wi-Fi, enabling cloud quantum computing and unbreakable privacy for everyone.

Conclusion: The Entangled Web Ahead

The quantum internet is not just a faster web—it's a **new communication paradigm**, rooted in the laws of physics. By enabling secure, distributed quantum computation and communications, it will reshape **cybersecurity, science, and geopolitics**.

Like the classical internet before it, the quantum internet will unlock applications we can't yet imagine—and connect the quantum future across the globe.

Chapter 52

Ethical Considerations in Quantum Development

Quantum computing, with its transformative potential, is poised to disrupt nearly every aspect of modern life—from cryptography and defense to drug discovery and artificial intelligence. But as with any powerful technology, its development carries **serious ethical responsibilities**.

At this early stage, stakeholders have a rare opportunity: to **embed ethics into the foundations** of the quantum ecosystem—before it becomes deeply entrenched in global systems and power structures.

This chapter explores key ethical dimensions of quantum technology, including **security, access, equity, governance, and unintended consequences**, offering a roadmap for responsible innovation.

Why Ethics Matters in Quantum Tech

Ethics in quantum development isn't just about what we *can* do—it's about what we *should* do. Quantum technologies introduce:

- **Unprecedented computational power**
- **Global security disruption potential**
- **Opportunities for misuse in surveillance, warfare, and inequality**

Because these impacts may be **asymmetrical and irreversible**, it's critical to apply foresight and deliberation now—not after deployment.

1. Cryptographic Disruption and Global Security

Quantum computers could break today's widely used cryptographic standards (e.g., RSA, ECC), endangering:

- Financial transactions
- Private communications
- National security systems
- Digital identities

Ethical risks:

- Exploiting quantum advantage before defenses are upgraded
- Creating geopolitical imbalances in intelligence capabilities
- Covert surveillance of protected or private data

Mitigation:

- Promote **transparent post-quantum cryptography standards**.
- International treaties on responsible quantum use.
- Public-private collaborations on secure infrastructure upgrades.

2. Inequitable Access and Digital Divides

Quantum computing may become concentrated among a few elite players—governments, tech giants, or high-income countries—while others are left behind.

Risks:

- Widening the **technological gap** between rich and poor nations
- Limiting access to quantum-enhanced science or medicine
- Reinforcing monopolies and reducing academic freedom

Mitigation:

- Develop open-source quantum tools and learning resources.
- Support **global quantum education initiatives**.
- Promote **public funding** and international collaboration for equitable access.

3. Environmental and Resource Costs

Quantum hardware—particularly superconducting and cryogenic systems—requires significant energy and rare materials. As systems scale:

- **Cryogenic cooling** and specialized manufacturing may strain energy grids.
- **E-waste** and toxic materials could become problematic.

Ethical imperative:

- Design for **energy efficiency and recyclability**.
- Conduct **lifecycle assessments** for quantum hardware.

- Set environmental **impact standards** for quantum labs and datacenters.

4. Military Use and Dual-Use Dilemmas

Quantum sensors, secure comms, and code-breaking capabilities have clear military applications.

Concerns:

- Accelerated **arms races** and autonomous weapon systems
- Uneven power dynamics and risk of cyberwar escalation
- Secret development of capabilities for offensive use

Ethical response:

- Support **international dialogue** on dual-use quantum technologies.
- Create oversight bodies for **ethical export and use**.
- Build **whistleblower protections** for quantum researchers.

5. Surveillance, AI, and Civil Liberties

Quantum-enhanced AI could speed up the analysis of massive datasets, raising risks in:

- Facial recognition
- Behavioral prediction
- Real-time location tracking

When deployed without oversight, these technologies threaten **privacy, consent, and autonomy**.

Ethical stance:

- Require **ethical impact assessments** for quantum-AI systems.
- Uphold **digital rights and transparency** in quantum-powered analytics.
- Advocate for **human-centered design** in AI and quantum fusion.

6. Governance, Transparency, and Open Science

Quantum development currently lacks unified **ethical governance structures**.

Risks:

- Proprietary secrecy in algorithmic decision-making
- Limited public understanding or input
- Inconsistent ethical norms across countries and corporations

Solutions:

- Encourage **ethical review boards** for major quantum projects.
- Build **standards for transparency and reproducibility**.
- Promote **multilateral treaties** on quantum R&D ethics.

7. Existential Risks and Long-Term Responsibility

While speculative, some experts warn that quantum computing may catalyze:

- **Runaway AI systems** enabled by accelerated training
- Breakdown of **global cybersecurity norms**
- Displacement of high-trust digital institutions

Ethical foresight demands:

- Investment in **quantum safety research**
- Scenario planning for **unintended consequences**
- Multi-stakeholder governance models with foresight capacity

8. Inclusion, Education, and Workforce Diversity

Quantum science is still dominated by a narrow demographic, risking **blind spots** in problem framing and impact forecasting.

Remedies:

- Fund quantum education in **underrepresented regions and groups**
- Design inclusive research teams and leadership
- Center equity in national quantum initiatives

Ethical quantum development means **who builds it matters** just as much as what is built.

Toward an Ethical Framework for Quantum Technology

Principle	Description
Transparency	Open algorithms, public disclosures
Justice	Equitable access and impact
Privacy	Respect and enforce digital rights
Stewardship	Design for long-term sustainability
Accountability	Clear oversight and governance mechanisms
Inclusivity	Representation and participation from all groups

These principles can guide governments, companies, and research institutions toward **responsible innovation**.

Conclusion: Building a Quantum Future With Integrity

Quantum technology is powerful, complex, and still in its infancy. This makes it the perfect moment to shape its trajectory responsibly.

The decisions we make now—about who controls it, who benefits, and how it's used—will echo for generations. The ethical path forward demands **transparency, collaboration, and humility** as we unlock one of the most powerful tools humanity has ever conceived.

The future of quantum must be not only intelligent—but just.

Chapter 53

Interdisciplinary Research in Quantum Fields

Quantum computing is not simply a branch of physics—it is a **fusion of disciplines**. Its success depends on collaborative contributions from fields as varied as **physics, computer science, electrical engineering, materials science, mathematics, chemistry, and philosophy**.

The inherently complex and novel nature of quantum phenomena means that no single discipline can tackle its challenges alone. Whether developing better qubits, new algorithms, or ethical guidelines, progress is driven by **interdisciplinary research** that blends deep domain knowledge with quantum-specific insight.

This chapter explores how these diverse fields converge—and why such collaboration is the backbone of quantum innovation.

The Need for Interdisciplinary Collaboration

Quantum systems combine **physical laws, computational abstraction**, **information theory**, and **engineering constraints**. Each field contributes essential tools:

- **Physics**: Describes the quantum phenomena enabling qubits.
- **Computer Science**: Frames algorithms and complexity theory.
- **Mathematics**: Provides the linear algebra and group theory that underpin qubit dynamics.

- **Engineering**: Builds and scales hardware with high precision and stability.
- **Chemistry**: Supplies use cases like molecular modeling and drug discovery.
- **Ethics and Philosophy**: Ask how quantum affects society, truth, and identity.

Siloed efforts cannot capture the full complexity or opportunity of quantum systems.

Physics: The Bedrock of Quantum Understanding

Quantum computing is grounded in physical principles:

- **Superposition**, **entanglement**, and **measurement theory** drive the logic.
- **Condensed matter** and **atomic physics** help build physical qubits.
- Quantum field theory shapes long-term applications like quantum gravity and cosmology.

Experimental physicists work closely with engineers to build viable devices, while theorists develop new models for error correction and topological computation.

Computer Science: Framing the Quantum Problem

CS brings structure and abstraction:

- **Quantum complexity classes** (e.g., BQP, QMA) define what's computable.

- **Quantum algorithms** like Shor's and Grover's come from algorithmic insight.
- **Programming languages** and **compilers** translate logic into gates and circuits.

It also leads the charge in:

- **Verification and debugging tools**
- **Hybrid quantum-classical architectures**
- **Quantum machine learning frameworks**

Mathematics: The Language of Quantum Theory

Advanced mathematics is woven into every layer:

- **Linear algebra**: State vectors, operators, unitary transformations.
- **Abstract algebra and group theory**: Symmetries in quantum gates and particles.
- **Information theory**: Entropy, fidelity, quantum channel capacities.
- **Topology**: Foundational to topological quantum computing and error codes.

Mathematics enables **rigorous reasoning** about abstract quantum behaviors, forming the connective tissue between theory and practice.

Engineering: Turning Theory into Machines

Engineering disciplines bring ideas to life:

- **Electrical Engineering**: Microwave control, signal processing, circuit design.
- **Mechanical Engineering**: Vibration isolation, cryogenic system design.
- **Materials Science**: Development of superconducting substrates, ion traps, or photonic chips.
- **Control Systems**: Feedback loops for dynamic calibration and error suppression.

These experts build quantum processors, **design packaging and control systems**, and prototype scalable platforms.

Chemistry and Biology: Application-Driven Collaboration

Many early quantum applications are in **quantum chemistry**, where interdisciplinary work is essential:

- Chemists define **problem Hamiltonians** for molecules.
- Quantum physicists simulate these using **variational algorithms**
- Computer scientists optimize circuit depth and resource use.

Similarly, **biology and medicine** are emerging areas:

- Protein folding, drug docking, and genomics involve **quantum-accelerated data analysis**.
- Hybrid quantum-AI systems may one day uncover new biological mechanisms.

Philosophy and Ethics: Interpreting Impact

Quantum mechanics has long raised **philosophical questions** about reality, observation, and information. Now, quantum computing prompts ethical questions:

- What is knowledge when truth becomes probabilistic?
- How do we regulate something we don't yet fully understand?
- Who owns quantum breakthroughs—and who is responsible for their consequences?

Philosophers and ethicists help society **grapple with meaning, responsibility, and governance**.

Education: Creating the Interdisciplinary Quantum Workforce

Training future quantum professionals requires **breaking down academic barriers**. Leading institutions now offer:

- **Quantum Information Science degrees** combining physics, CS, and engineering.
- **Joint research labs** uniting theorists and experimentalists.
- **Quantum bootcamps and upskilling programs** for industry professionals.

This education model is crucial for cultivating **"quantum generalists"** who understand the interfaces between disciplines.

Collaborative Platforms and Ecosystems

Many global initiatives promote interdisciplinary research:

Initiative	Focus
QED-C (USA)	Public-private quantum research ecosystem
QuTech (Netherlands)	Integrated physics-engineering projects
Quantum Flagship (EU)	Cross-national, cross-disciplinary R&D
NSF QLCI	US-based centers uniting physics, CS, and ethics
Google Quantum AI, IBM Quantum, Microsoft Station Q	Blend research and application development

These hubs produce breakthroughs by creating **shared languages, priorities, and tools** across domains.

Emerging Interdisciplinary Research Themes

- **Quantum cyber-physical systems**: Integrating quantum sensing into autonomous vehicles and robots.
- **Quantum economics**: Using quantum methods to model uncertainty in markets.
- **Quantum linguistics**: Exploring quantum-like models in human language and cognition.

- **Quantum sociology and geopolitics**: Investigating quantum's influence on power structures and global diplomacy.

Each of these frontiers requires **deep integration of disciplines** that once rarely interacted.

Conclusion: The Fusion Mindset

The future of quantum technologies will not be built in a physics lab alone—or in a server room, or in a policy office. It will be created in **collaborative spaces where multiple disciplines converge**, where physicists speak code, engineers read papers on entanglement, and philosophers advise on data policy.

Interdisciplinary research is not just beneficial in quantum—it is foundational.

As quantum systems become more integrated into society, the need for **collaboration, diversity of thought, and boundary-crossing innovation** will only grow.

Chapter 54

Quantum Workforce and Education

Quantum computing is transitioning from academic novelty to industrial reality. But this transition hinges on a critical ingredient: **people**. Without a skilled, diverse, and scalable workforce, even the most advanced quantum hardware and algorithms will sit idle.

As the field grows beyond elite research labs and government programs, the challenge becomes clear: How do we **educate, train, and empower a global quantum workforce**?

This chapter examines current gaps, emerging education strategies, and the interdisciplinary nature of preparing the world for the **quantum revolution**.

Why the Quantum Talent Gap Exists

Quantum computing sits at the intersection of multiple deep domains—physics, computer science, engineering, and mathematics. Historically, training in these areas has been **highly siloed**.

Challenge	Impact
Narrow academic pipelines	Too few with cross-domain fluency
Limited industry training	Engineers lack quantum-specific exposure

Low K-12 awareness	Students unaware of quantum career paths
Diversity barriers	Women and minorities underrepresented

To fill thousands of quantum jobs in the coming decade, **new educational models** must scale quickly and inclusively.

What Skills Define a Quantum-Ready Workforce?

Quantum roles require a blend of **domain expertise and hands-on skills**. The spectrum includes:

Core Competencies

- Linear algebra and probability theory
- Quantum mechanics and information theory
- Quantum gates, circuits, and algorithms
- Familiarity with SDKs (Qiskit, Cirq, PennyLane, etc.)

Applied Skills

- Circuit optimization and error mitigation
- Hardware control and pulse programming
- Quantum machine learning and VQE/QAOA usage
- Hybrid quantum-classical programming

Professional Skills

- Scientific communication

- Team-based problem solving
- Ethical decision-making in emerging tech

Education Pathways: From High School to PhD

Level	Focus Area	Example Programs
K–12	Basic quantum concepts, STEM curiosity	Quantum for All, QuSTEAM outreach
Undergraduate	Multidisciplinary curriculum in STEM	MIT, University of Waterloo, Caltech
Graduate	Advanced research in quantum physics/CS	Oxford, Harvard, ETH Zurich, TUM
Bootcamps	Rapid upskilling for industry professionals	Qiskit Global Summer School, QWorld
Online Courses	Accessible global learning	edX, Coursera, Brilliant.org, QuEra

Programs are expanding beyond top-tier institutions to **community colleges, online universities, and workforce boards**.

Emerging Educational Models

1. **Quantum Literacy Initiatives**
    - Introduce core ideas like superposition and entanglement early.

- Use visual and gamified tools (e.g., Quirk, Hello Quantum).

2. **Interdisciplinary Curricula**
    - Combine CS, physics, ethics, and business.
    - Develop joint degrees in Quantum Information Science and Engineering (QISE).
3. **Industry-Academia Partnerships**
    - Internships, co-ops, and real-world project labs with companies.
    - Examples: IBM Q Network, Microsoft Q#, Google Quantum AI internships.
4. **Quantum Labs and Simulators**
    - Cloud-based access to real quantum computers.
    - Encourage experimentation and hands-on learning.
5. **Credentialing and Badging**
    - Quantum certification paths to validate industry skills.
    - Examples: Qiskit Advocate, D-Wave Developer, edX Quantum Certificate.

Roles in the Quantum Ecosystem

Role	Primary Skills	Backgrounds Welcome
Quantum Algorithms Developer	Math, linear algebra, circuit design	CS, math, physics
Quantum Software Engineer	SDKs, compilers, API design	CS, EE, software engineering
Quantum Physicist	Hardware modeling, decoherence	Physics, applied math
Quantum Hardware Engineer	Materials, control systems	EE, materials science, optics
Quantum Product Manager	Domain knowledge, user experience	CS, business, UX, engineering
Quantum Educator/ Trainer	Communication, curriculum design	Education, physics, CS

These roles are increasingly filled by **hybrid experts** trained across multiple domains.

Equity and Inclusion in Quantum Education

To ensure a **just and innovative ecosystem**, quantum education must prioritize:

- Outreach to **underserved communities**
- Support for **first-generation learners**
- Scholarships and mentoring for **women and minorities**

- Localization of content in **multiple languages and cultures**

Programs like **Unconventional Computing Lab (UWE)** and **Qubit by Qubit** lead efforts to democratize access.

The Role of Governments and Foundations

Public and nonprofit investment is key to scaling quantum education:

Organization	Initiative
NSF (USA)	Quantum Leap Challenge Institutes
DOE/NIST	QIS education grants and fellowships
European Quantum Flagship	EQTC and pan-European education efforts
UNESCO	Global quantum literacy and outreach
Canada's Quantum Strategy	National workforce funding and training hubs

Governments increasingly recognize quantum education as **strategic infrastructure**.

Metrics for Success

To track progress, we need to monitor:

- Growth in **quantum-capable graduates**
- Job placement in quantum roles
- Increase in **women and underrepresented minorities**
- Participation in **public/online quantum courses**
- Industry feedback on workforce readiness

Conclusion: A Quantum-Ready World Starts with Education

Quantum computing has the potential to transform how we solve problems, build businesses, and understand the universe. But this promise will only be realized if **millions of people, not just a few thousand, are equipped to understand and use it**.

By investing in education today—from middle schools to industry bootcamps—we're building the foundation for a **resilient, equitable, and innovative quantum future**.

The age of quantum belongs to the prepared—and that preparation starts now.

Section 10: Visionary Concepts and Theoretical Horizons

Chapter 55

Quantum Gravity and Black Hole Information Paradox

Quantum computing may revolutionize industries and science, but it also opens doors to **cosmic questions**—none more mysterious than the nature of **black holes**, and how they relate to **quantum gravity**.

Black holes sit at the intersection of **general relativity** and **quantum mechanics**, two foundational but seemingly incompatible theories. Understanding how gravity works in the quantum realm—and whether black holes destroy information—has become one of the deepest problems in physics.

This chapter explores how quantum computing and information theory may help resolve the **black hole information paradox**, and how these ideas point toward a long-sought theory of **quantum gravity**.

The Clash Between Gravity and Quantum Theory

Einstein's **general relativity** describes gravity as the curvature of spacetime—smooth, continuous, and deterministic.

Quantum mechanics, on the other hand, is **probabilistic**, **discrete**, and dominated by uncertainty.

For most phenomena, these frameworks coexist peacefully. But near black holes and at the **Planck scale**, both are necessary—yet **conflict**.

What Is the Black Hole Information Paradox?

In the 1970s, **Stephen Hawking** discovered that black holes aren't truly black—they emit radiation due to quantum effects near the event horizon. This radiation is **thermal** and appears to contain no information about what fell in.

But here's the paradox:

1. **Quantum mechanics** says information is never lost (unitary evolution).
2. **Hawking radiation** suggests that black holes evaporate and destroy information.
3. If both are true, **something must break**.

This is the **black hole information paradox**—a tension between the laws of gravity and quantum theory.

Why This Matters for Quantum Science

This isn't just a philosophical issue. It challenges:

- The **foundations of quantum mechanics** (unitarity, reversibility).
- The **consistency of general relativity** at high energies.
- The **definition of entropy** and the limits of quantum computation.

Resolving it could unlock:

- A full theory of **quantum gravity**.
- A deeper understanding of **space, time, and information**.

- New models of **holography and quantum entanglement**.

Hawking Radiation and the Loss of Information

Hawking showed that black holes emit radiation with a temperature proportional to their surface gravity. Over time, they **evaporate**.

But this radiation appears **random**—like pure heat. No trace of the **information about infalling matter** seems to remain.

If the black hole disappears, **where does the information go?**

- Into the radiation?
- Into a hidden remnant?
- Or is it truly lost?

Proposed Resolutions to the Paradox

1. **Information is encoded in Hawking radiation**
    - Subtle quantum correlations ("scrambling") might preserve information.
    - Supported by **unitarity** and **Page curve** calculations.
2. **Black holes have "soft hair"**
    - Information stored on the event horizon's quantum states.
3. **Holographic Principle**

- Information inside a volume can be encoded on its boundary.
- Leads to the **AdS/CFT correspondence**: a duality between gravity in spacetime and quantum field theory without gravity.

4. **Firewalls**
    - Radical idea: to preserve unitarity, the event horizon is a wall of high-energy particles that breaks general relativity.
5. **Final state proposal**
    - Black holes project information out at the end of evaporation—possibly preserving unitarity.

Quantum Information Theory Enters the Scene

Recent advances suggest that **black holes behave like the most efficient information scramblers in nature**.

- Entanglement plays a central role in understanding the **black hole interior**.
- Tools like **quantum teleportation** and **error correction** offer models for black hole dynamics.
- **Quantum circuit complexity** has been proposed as a measure of gravitational entropy.

The Page Curve: A Clue to Unitarity

Don Page predicted a shape for entropy over time if information is preserved:

- At first, Hawking radiation looks random, and entropy rises.
- After the "Page time," correlations increase, and entropy **falls**.
- Total entropy returns to zero if evaporation is unitary.

Recent work using **quantum extremal surfaces** and **replica wormholes** supports the **Page curve**, aligning quantum gravity with unitarity.

How Quantum Computers Help

Quantum computers allow simulation of **toy models of black holes**:

- SYK models (Sachdev -Ye- Kitaev) mimic chaotic scrambling.
- Tensor networks model **holographic spacetime geometry**.
- Quantum machine learning explores entanglement and emergent geometry.

These experiments can test ideas that were once purely theoretical.

Toward a Theory of Quantum Gravity

To resolve the paradox fully, we need a **unified theory** that:

- Quantizes gravity consistently.
- Describes how space and time emerge from quantum states.
- Preserves information and causality.

Leading candidates:

- **String theory** and **AdS/CFT duality**
- **Loop quantum gravity**
- **Holographic quantum error correction codes**

Quantum information is increasingly central to all of them.

Conclusion: Black Holes as the Ultimate Quantum Lab

The black hole information paradox once seemed like a thought experiment. Now, it is a crucible for the unification of physics.

By linking **entanglement, entropy, geometry, and computation**, black holes may teach us how to **rebuild spacetime from quantum rules**.

The fusion of quantum computing and gravitational physics doesn't just solve a paradox—it may reveal how the universe is woven together.

Chapter 56

Time Crystals and Quantum Phases of Matter

Time Crystals and Quantum Phases of Matter

In classical physics, matter is categorized by its **phases**—solid, liquid, gas, plasma—based on how atoms arrange in space. In quantum physics, however, new kinds of phases emerge, often governed not by temperature or pressure but by **symmetry, entanglement, and topology**.

Among these, one concept has captured scientific imagination like no other: the **time crystal**.

Time crystals are a new phase of matter that exhibit **periodic motion in time**, even in their ground state—without absorbing energy. Like spatial crystals break **translational symmetry in space**, time crystals break **translational symmetry in time**.

This chapter explores how time crystals were theorized, created, and what they mean for **quantum systems, nonequilibrium physics, and future technologies**.

What Is a Phase of Matter in Quantum Systems?

In quantum mechanics, phases of matter go far beyond solid, liquid, and gas. Examples include:

- **Superconductors**
- **Bose–Einstein condensates**

- **Topological insulators**
- **Fractional quantum Hall states**

These phases are defined by:

- **Long-range quantum order**
- **Entanglement patterns**
- **Symmetry breaking or protection**

Time crystals fall under the umbrella of **nonequilibrium quantum phases**.

What Is a Time Crystal?

A **time crystal** is a system that:

1. **Exhibits periodic behavior in time**, like ticking, without external input.
2. **Does not absorb energy**—it's in its **lowest energy state**.
3. **Breaks time-translation symmetry**: The laws of physics are the same at all times, but the system chooses a specific rhythm.

This motion is **not due to energy oscillations** (like a pendulum), but an intrinsic quantum effect—like a **"ground-state clock."**

Theoretical Origins: Frank Wilczek's Proposal

In 2012, Nobel laureate **Frank Wilczek** proposed time crystals as a novel symmetry-breaking phase in quantum systems. Initially, it sparked controversy—some argued such motion would **violate conservation laws**.

Later, physicists refined the idea to focus on **discrete time crystals (DTCs)**:

- Systems driven periodically (like a strobe light)
- Respond with a **different periodicity** (e.g., every 2 ticks instead of every 1)
- Stabilized by **many-body localization** or **long-range order**

Key Properties of Time Crystals

Property	Description
Temporal periodicity	Repeats motion at fixed intervals
No energy input required	Exists in or near ground state
Long-range quantum order	Nonlocal entanglement across the system
Robust to decoherence	Protected by disorder or interaction
Symmetry breaking	Time translation symmetry spontaneously broken

Time crystals **respond periodically**, but not because of classical oscillations—they're **topological or interaction-stabilized** patterns in time.

First Experimental Realizations

In 2017, two independent groups realized **discrete time crystals**:

1. **Google + Princeton + UC Berkeley (in a chain of trapped ions)**
   - Used laser pulses and interactions to produce subharmonic responses.
2. **Harvard + MIT + Maryland (with NV centers in diamond)**
   - Observed stable temporal oscillations in a disordered system.

These systems **oscillated in a stable rhythm** not dictated by the driving field—a key signature of time-crystalline order.

Quantum Simulators and Time Crystals

Quantum simulators like **superconducting qubit arrays** and **trapped ions** are ideal for studying time crystals because they allow:

- Precise control over interactions
- Isolation from thermal environments
- Programmable disorder or periodic driving

In 2021, **Google's Sycamore quantum computer** observed time crystal behavior in a programmable quantum system—an experimental milestone.

Time Crystals vs. Classical Oscillators

Feature	Time Crystal	Classical Oscillator
Energy required	No (ground state)	Yes (external energy input)
Source of motion	Spontaneous symmetry breaking	Newtonian force
Robust to perturbations	Yes (in ideal systems)	Often not
Role of quantum effects	Essential	Absent or negligible

Time crystals represent a **genuinely new form of organized motion**, governed by **quantum coherence and interactions**.

Applications and Implications

Though still nascent, time crystals may enable:

- **Quantum memories**: Information encoded in stable temporal order
- **Precision metrology**: Oscillations locked to quantum interactions
- **Quantum networks**: Synchronizing qubits across nodes
- **New phases in condensed matter**: Beyond equilibrium thermodynamics

Time crystals challenge the traditional notion that order arises from **energy minimization**. Instead, they show that **order can emerge in time** under periodic driving and isolation.

Related Quantum Phases and Concepts

Concept	Relation to Time Crystals
Many-body localization	Protects time crystal order in disordered systems
Floquet engineering	Creates time-periodic Hamiltonians
Topological phases	Both exhibit protected edge or dynamical states
Prethermalization	Time crystals can exist in prethermal regimes
Spin glasses	Disorder-induced memory, like time crystals

These concepts reflect the **growing diversity of quantum phases** —not just of matter, but of **dynamics**.

Challenges in Time Crystal Research

- Maintaining **coherence over time**
- Engineering **long-range interactions**

- Distinguishing true time-crystal behavior from transient oscillations
- **Scaling** up to larger and more complex quantum systems

Despite these, experimental control and theoretical tools are rapidly advancing.

Conclusion: A New Kind of Order in the Universe

Time crystals stretch our understanding of symmetry, motion, and equilibrium. They don't merely oscillate—they **redefine time's role** in quantum systems.

By exploring how quantum order can arise in **space *and* time**, physicists are opening new pathways to **quantum control, computation, and fundamental understanding**.

As quantum technologies grow more complex, time crystals remind us that **entirely new phases of reality are still waiting to be discovered**.

Chapter 57

Quantum Supremacy: Definitions and Controversies

Quantum supremacy is a term that once promised a clear threshold in computational history—the point where a quantum computer outperforms the best classical supercomputers on a specific task. It was hailed as the moon landing of quantum computing.

Yet, as with many scientific landmarks, the reality is far more nuanced. The definition, interpretation, and implications of quantum supremacy have sparked **technical, philosophical, and ethical debates**.

This chapter explores what quantum supremacy means, how it has (and hasn't) been achieved, and why it remains one of the most **important—and controversial—concepts** in modern quantum science.

What Is Quantum Supremacy?

Originally coined by physicist **John Preskill** in 2012, the term *quantum supremacy* refers to:

"The point at which a quantum computer can perform a computational task that is infeasible for any classical computer."

This does **not** require solving a useful real-world problem. Rather, it focuses on any task—however abstract—that clearly demonstrates **quantum computational advantage**.

Why Is It a Big Deal?

Achieving quantum supremacy is symbolic of:

- **Quantum computers doing something classically impossible**.
- Validation that **quantum mechanics enables superior computation**.
- A shift from quantum theory to **quantum reality**.

It serves as proof-of-principle that quantum processors have crossed a fundamental threshold—even if only for narrow tasks.

Google's 2019 Supremacy Claim

In 2019, Google announced that its 53-qubit **Sycamore processor** had achieved quantum supremacy by:

- Sampling from the output of a **random quantum circuit**.
- Taking about **200 seconds** for a task estimated to require **10,000 years** on a classical supercomputer.

Published in *Nature*, the experiment sparked global headlines—but also pushback.

Controversies and Criticisms

❗ IBM's Rebuttal

Shortly after Google's announcement, **IBM argued** that their Summit supercomputer could simulate the same task in **2.5 days**, not 10,000 years—thus challenging the claim of "supremacy."

❗ Usefulness vs. Theatrics

Critics noted:

- The task had **no practical utility** (random circuit sampling).
- The bar for "infeasibility" was **artificially set**.
- Supremacy sounded like a **marketing stunt** more than a scientific breakthrough.

❗ Terminology Concerns

The word "supremacy" carries political and social overtones that many researchers consider inappropriate. Alternative terms proposed include:

- **Quantum advantage**
- **Quantum primacy**
- **Quantum computational superiority**

Beyond Google: Other Attempts and Progress

Institution	Achievement/Focus
USTC (China)	Gaussian boson sampling with photonic chips
IBM	Scalable superconducting circuits, but no supremacy claim

Xanadu	Boson sampling with photonic qubits
IonQ	Focus on algorithmic quantum advantage

While Google's 2019 paper remains a milestone, **subsequent efforts** are reframing the goal to be more about **useful quantum advantage**, not just raw supremacy.

The True Challenges of Supremacy

Achieving quantum supremacy is not just about building bigger machines—it requires addressing:

- **Noise and error rates**: Current devices are NISQ (noisy intermediate-scale quantum).
- **Benchmarking fairness**: Classical algorithms constantly improve.
- **Replicability**: Supremacy demonstrations are hard to independently verify.

Moreover, **quantum supremacy does not mean general-purpose quantum computing** is here—it only confirms **one type of superiority on one task**.

Useful Quantum Advantage vs. Supremacy

Supremacy	Useful Advantage
Abstract or synthetic problem	Real-world application

One-off demonstration	Repeatable, reliable results
Little relevance to industry	Tangible value for businesses
Symbolic milestone	Operational competitiveness

The field is shifting from supremacy toward **scalable, practical quantum value**.

The Role of Classical Algorithms in the Debate

Ironically, every quantum supremacy claim pushes classical computing **to get better**.

- Post-Google, new **tensor network methods** emerged for circuit simulation.
- Quantum supremacy races have become **mutually beneficial for both fields**.

This raises philosophical questions:

- Can supremacy ever be truly "settled," or will classical methods always catch up?
- Does this imply a moving goalpost for the entire concept?

Implications for Industry and Society

- Supremacy is a **PR tool**, helping secure funding and media attention.

- It boosts **investor confidence**, even if the benefits are still distant.
- But overstating it risks **damaging credibility**, especially if real-world advantage takes years to arrive.

A Landmark With Limits

Quantum supremacy is best seen as a **scientific inflection point**, not a commercial breakthrough.

It confirms that quantum machines can outperform classical ones—in very narrow ways. But the road to **general-purpose, fault-tolerant, and useful quantum computing** remains long.

Supremacy may not be the end goal—but it's an important signpost that we are no longer just theorizing about quantum computation. We're building it.

Chapter 58

The Many-Worlds Interpretation in Computing

Quantum computing thrives on the weirdness of quantum mechanics—superposition, entanglement, and interference. But what *do* these phenomena *mean*?

One particularly provocative framework is the **Many-Worlds Interpretation (MWI)**. First proposed by Hugh Everett in 1957, it suggests that **all possible outcomes of quantum events actually occur**, each in a parallel "branch" of reality.

While MWI is mostly a philosophical interpretation of quantum mechanics, it raises fascinating questions for computing:

- Are quantum algorithms literally "computing in parallel universes"?
- Can MWI help explain quantum speedup?
- Does it reframe our understanding of information, causality, and identity?

This chapter investigates how the **Many-Worlds lens** helps us think differently about **quantum computation**, even if it's not required for practical execution.

What Is the Many-Worlds Interpretation?

In the traditional **Copenhagen interpretation**, measurement causes a wave function to "collapse" into one outcome, chosen probabilistically.

In contrast, MWI posits:

- **No collapse occurs.**
- **All possible outcomes happen**, each in its own "world" or branch.
- The observer becomes entangled with the result, splitting into multiple versions.

Thus, quantum measurement doesn't select a single future—it **branches into many**.

Why This Matters for Quantum Computing

Quantum computers rely on **superposition**—qubits exist in multiple states simultaneously. During computation, quantum circuits explore **interference between many possibilities**.

MWI proposes a striking way to visualize this:
Each computational path exists in a **separate universe**, and quantum interference is **communication across these universes**, selecting outcomes through constructive and destructive overlap.

Is Quantum Speedup Due to Parallel Universes?

Popular science often suggests that quantum computers "try all answers at once in parallel worlds."

This metaphor is **intuitive—but oversimplified**. Here's why:

Claim	Reality
Qubits explore all possibilities	True (via superposition)

Each result is in a separate world	MWI-specific interpretation
Speedup comes from many-worlds	Not proven; depends on interference, not branching
All answers are tested in parallel	Only the final result is extracted—**not all paths**

Quantum computing exploits **wave function structure and interference**, regardless of whether you believe in MWI, Copenhagen, or other interpretations.

Quantum Interference: The Key Mechanism

MWI frames interference as **interactions among parallel branches**:

- In Grover's algorithm, incorrect paths cancel each other across universes.
- In Shor's algorithm, periodicity emerges from harmonized branches.

From this view, quantum computation becomes a **global coordination of amplitudes across possible realities**.

Does MWI Explain the Limits of Quantum Computing?

While MWI offers a compelling visual model, it **does not increase computational power**.

Quantum computers can't:

- Solve NP-complete problems in polynomial time.

- Explore *all* solutions with equal resolution.
- Extract multiple results at once.

The interference pattern is **carefully engineered**, and measurement yields just **one answer**.

Philosophical Implications for Computing

MWI raises deep questions about:

- **What computation *is***—if all outcomes exist, is "solving" just selecting the right branch?
- **What is identity?**—Does running a quantum algorithm split the programmer across realities?
- **Information conservation**—If no outcome is destroyed, does information ever get erased?

While speculative, these ideas **force us to re-express computing in probabilistic and entangled terms**, not binary logic.

MWI vs Other Interpretations in Computation

Interpretation	Collapse?	Multiverse?	Role in Quantum Computing
Copenhagen	Yes	No	Traditional, widely taught
Many-Worlds	No	Yes	Popular among theorists, no operational difference

de Broglie–Bohm	No	No	Hidden variables guide evolution
QBism	No	No	Personalist Bayesian interpretation

All interpretations **make identical predictions**. Their role in computing is **conceptual**, not functional.

The Role of MWI in Quantum Algorithms Education

Some educators use MWI metaphors to explain:

- How quantum computers explore multiple paths.
- Why interference matters.
- What happens during entanglement and measurement.

This can help **demystify algorithms**—but must be paired with rigorous understanding of **amplitude manipulation**, not naive parallelism.

In Fiction and Culture

MWI has inspired:

- **Science fiction (e.g., "Everything Everywhere All at Once")**
- **Quantum-themed games**
- **Speculations about quantum consciousness**

In computing, it fuels creative thinking but should be distinguished from **scientific engineering practice**.

Conclusion: Many Worlds, One Computer

Whether you embrace it as truth or metaphor, the Many-Worlds Interpretation invites us to think **beyond linear causality**.

In quantum computing, MWI illustrates the elegance of **interference**, the mystery of **measurement**, and the richness of **possibility**.

Yet, at the end of the computation, there's only one result.
And in our world, that's the one we see.

Chapter 59

Decoherence and the Quantum-Classical Boundary

Quantum computers derive their power from principles like superposition and entanglement—behaviors that are profoundly unlike anything in classical physics. But this power is fragile. When quantum systems interact with their environment, they quickly lose their coherence, and the system begins to behave **classically**.

This loss of quantum behavior is called **decoherence**. It's not just a technical hurdle; it's one of the fundamental processes that defines the boundary between the quantum and classical worlds.

This chapter explores what decoherence is, how it undermines quantum computing, and how researchers are fighting to preserve quantum states long enough to compute something useful.

What Is Decoherence?

Decoherence is the process by which a **quantum system loses its quantum properties** due to interaction with its surrounding environment.

In simple terms:

- A qubit in a superposition of $|0\rangle$ and $|1\rangle$ will collapse into either state when it becomes entangled with the environment.

- This collapse isn't a measurement per se—it's a **loss of phase information**, making the system **indistinguishable from a classical mixture**.

Decoherence vs. Measurement

Process	Description	Outcome
Measurement	Intentional interaction with apparatus	Collapse to one outcome
Decoherence	Uncontrolled environmental influence	Loss of coherence without collapse (but appears classical)

Decoherence **mimics measurement** from an outside observer's perspective, but it's fundamentally a **dynamical loss of quantum correlation**.

Sources of Decoherence in Quantum Hardware

Hardware Platform	Common Decoherence Sources
Superconducting Qubits	Charge noise, flux noise, dielectric loss
Trapped Ions	Magnetic field fluctuations, motional heating

Photonic Qubits	Scattering, phase drift, loss in fiber
Spin Qubits	Spin–spin interactions, thermal noise

Environmental coupling, even at near-absolute-zero temperatures, introduces **irreversible noise**.

Coherence Times and Error Rates

Two key metrics define decoherence:

- **T_1 (relaxation time):** Time for energy to dissipate ($|1\rangle \rightarrow |0\rangle$).
- **T_2 (dephasing time):** Time for phase relationships to be lost.

A useful quantum computer requires **coherence times longer than the full computation**. Otherwise, quantum information degrades before it can be read.

The Quantum-Classical Boundary

Decoherence is the best explanation we have for **why we don't observe superpositions in the macroscopic world**.

- A cat can't be in a superposition because it interacts with trillions of air molecules, photons, and thermal vibrations every second.
- This interaction causes **decoherence times on the order of femtoseconds**—far too short to preserve quantum behavior.

Thus, decoherence **explains the quantum-to-classical transition** without invoking wave function collapse as a mystery.

Mathematical View: Density Matrices and Entanglement

In the **density matrix formalism**, decoherence manifests as the **off-diagonal terms** decaying over time:

$$\rho(t)=\begin{bmatrix} |\alpha|^2 & \alpha^*\beta e^{-t/T_2} \\ \alpha\beta^* e^{-t/T_2} & |\beta|^2 \end{bmatrix}$$

As time increases, the off-diagonal terms vanish → the state becomes **classically probabilistic**, not quantum coherent.

Fighting Decoherence: Mitigation Strategies

1. **Cryogenic Environments**
    - Cooling to millikelvin temperatures reduces thermal noise.
2. **Isolated Qubit Design**
    - Physical isolation via vacuum chambers or electromagnetic shielding.
3. **Dynamical Decoupling**
    - Use of fast pulse sequences to "refocus" qubit states and cancel errors.
4. **Quantum Error Correction**

    - Encoding logical qubits in ensembles of physical qubits to **detect and correct decoherence**.

5. **Topological Qubits (e.g., Majorana)**
    - Qubit states that are intrinsically protected from local noise.

Why Decoherence Matters in Practice

Decoherence limits:

- **Circuit depth** (how many operations can be done).
- **Algorithm fidelity** (likelihood of getting the right result).
- **Qubit scalability** (adding more qubits increases noise paths).

It's the central reason why **quantum hardware is still in the NISQ era** (Noisy Intermediate-Scale Quantum)—and not yet fault-tolerant.

Decoherence in Quantum Algorithms

Certain algorithms are **more sensitive** to decoherence:

- Shor's Algorithm: Requires long coherent time across many qubits.
- Grover's Algorithm: Tolerates some error due to its iterative nature.
- VQE/QAOA: Hybrid algorithms can correct in real time with classical feedback.

Understanding decoherence informs **which algorithms are viable on today's devices**.

Decoherence in Interpretation and Philosophy

Decoherence also reshapes foundational debates:

- **Many-Worlds Interpretation**: Decoherence defines branches but no collapse.
- **Copenhagen Interpretation**: Decoherence mimics collapse due to measurement.
- **Objective Collapse Theories**: Require decoherence plus an additional mechanism to choose a reality.

It helps **bridge the quantum and classical realms** without resorting to metaphysical assumptions.

Conclusion: Decoherence as the Gatekeeper of Quantum Power

Decoherence is both **a challenge and a key insight**:

- As a challenge, it's the main obstacle to reliable quantum computation.
- As an insight, it shows how quantum behavior gives way to classical physics.

Solving decoherence isn't just about engineering—it's about **understanding how reality transitions from the uncertain to the familiar**.

Preserving quantum coherence long enough to compute, communicate, and control information is what makes the dream of **usable quantum technology** real.

Chapter 60

The Road to Universal Quantum Computing

Quantum computing has made remarkable progress—from manipulating a few qubits in isolated labs to running real-time cloud experiments on quantum hardware. But we are still far from realizing the ultimate goal: a **universal, fault-tolerant quantum computer**.

A universal quantum computer is not just fast—it is **programmable, scalable, robust to errors, and capable of running any quantum algorithm** that can be defined in theory. Reaching this milestone would revolutionize fields from cryptography and drug discovery to materials design and beyond.

This chapter traces the roadmap toward universal quantum computing—highlighting breakthroughs, bottlenecks, and what it will take to cross the finish line.

What Is a Universal Quantum Computer?

A universal quantum computer is the quantum analog of a classical Turing machine: a device capable of executing **any computable function** using quantum gates, error correction, and measurement.

It can:

- Implement **arbitrary quantum circuits**
- Handle **entanglement, superposition, and interference**

- Correct for **decoherence and noise**
- **Scale up** without exponential cost

In essence, it's a general-purpose quantum processor—programmable like today's CPUs, but exponentially more powerful for certain problems.

Milestones on the Road to Universality

1. **Demonstration of basic quantum gates**
   ✓ Achieved in ion traps, superconducting circuits, and photonics.
2. **Entanglement and Bell inequality tests**
   ✓ Validated quantum non-locality in multi-qubit systems.
3. **Quantum supremacy (narrow task outperforming classical)**
   ✓ Claimed by Google in 2019.
4. **Quantum error correction (QEC)**
   ✓ Under active development; surface codes and bosonic codes showing promise.
5. **Logical qubits with sustained coherence**
   ⏳ In progress; most current qubits are physical (not error-protected).
6. **Fault-tolerant operations**
   ⏳ Requires hundreds to thousands of physical qubits per logical qubit.
7. **Scalable quantum architecture**
   ⏳ Still a bottleneck; integration and control complexity rising.

Key Components of a Universal Quantum Computer

Layer	Function
Physical qubits	Real-world implementation (e.g., superconducting, ion)
Quantum gates	Operations like Hadamard, CNOT, T, and rotations
Error correction codes	Detect and fix decoherence, gate errors
Quantum compiler	Translate algorithms into hardware-native instructions
Control systems	Pulse generation, calibration, feedback
Measurement and readout	Extract final classical result

Every layer must be **fault-tolerant and scalable** to reach universal capability.

The Challenge of Error Correction

Quantum systems are inherently **fragile**. Every logical qubit (used for computation) requires **many physical qubits**(used for error detection and correction).

For example:

- A single logical qubit using the **surface code** may need **1,000+ physical qubits** to be robust.
- Full-scale algorithms like factoring large numbers may require **millions** of qubits and billions of gate operations.

Without **error correction**, even the most powerful quantum chips will collapse under noise.

Hardware Approaches to Universality

Platform	Strengths	Limitations
Superconducting qubits	Fast, industrialized, scalable wiring	High error rates, cryogenic needs
Trapped ions	Long coherence times, precise gates	Slower gate speed, challenging scaling
Photonic qubits	Room temp, good for communication	Difficult nonlinear interactions
Topological qubits	Intrinsic error resilience	Still experimental (e.g., Majoranas)

No platform has yet emerged as dominant. Hybrid systems may combine strengths.

Software Challenges on the Path to Universality

1. **Quantum compilers**
    - Translate high-level code into gate sequences optimized for hardware.
2. **Quantum operating systems**
    - Schedule jobs, manage memory, route signals in multi-core systems.
3. **Benchmarking and verification**
    - Prove that a quantum device is executing algorithms correctly.
4. **Programming abstractions**
    - Make quantum software accessible to non-physicists via libraries like Qiskit, Cirq, PennyLane.

Toward Quantum Advantage at Scale

Universal computing doesn't mean every problem will run faster. But for some tasks, quantum computers offer **exponential advantage**:

- Factoring (Shor's algorithm)
- Unstructured search (Grover's algorithm)
- Quantum chemistry and physics simulation
- Quantum machine learning and optimization

These applications require **fault-tolerant, universal systems** to deliver real-world impact beyond today's noisy devices.

Timelines and Industry Perspectives

Prediction Source	Universal Quantum Computing Timeline
IBM, IonQ, Rigetti	10–15 years (aggressive)
Academic consensus	15–30 years (cautiously optimistic)
Skeptics	More than 50 years, if at all

Progress depends not just on qubits, but on **engineering scale, reproducibility, and funding continuity**.

Ethical and Societal Implications

As we approach universality, we must prepare for:

- **Cryptographic disruption**
- **Automation of scientific discovery**
- **Power imbalances** if only a few actors possess universal systems

Ethical foresight and governance must **grow in parallel with technical capabilities**.

Conclusion: A Long Road Worth Taking

The journey to universal quantum computing is one of the most complex engineering challenges of the 21st century. It requires:

- Breakthroughs in physics and materials,
- New paradigms in programming and control,
- Global collaboration across academia, industry, and government.

But the rewards—understanding nature more deeply, solving once-intractable problems, and expanding the boundaries of knowledge—make it a journey worth pursuing.

Universal quantum computing is not inevitable.
It's **a choice—and a challenge—that humanity must decide to meet**.

The Quantum Horizon

As we reach the end of this journey through the rich, complex, and often mysterious world of quantum computing, it is clear that we are witnessing the birth of a transformative era. What began as an abstract theory of subatomic behavior has grown into a multidisciplinary engine of discovery, pushing the boundaries of physics, computer science, mathematics, engineering, and even philosophy.

From the counterintuitive strangeness of superposition and entanglement to the rigorous demands of error correction and fault tolerance, the path toward scalable, universal quantum computers is as intellectually demanding as it is inspiring. Yet despite the obstacles, progress continues—driven by an extraordinary global community of researchers, educators, and innovators.

This book has aimed not only to explain the science behind quantum computing but also to place it in context: industrially, ethically, and philosophically. The road to quantum advantage is not merely about speed or power—it's about changing how we compute, how we understand reality, and how we prepare for a future shaped by quantum technologies.

As quantum systems become increasingly integrated into real-world applications—from cryptography and logistics to climate modeling and drug discovery—the need for broad awareness, education, and responsible innovation becomes greater than ever. The field is still young. The possibilities are vast. And the story is just beginning.

A Note of Gratitude

To you, the reader—thank you. Whether you are a researcher, student, policymaker, business leader, or curious mind, your time and attention mean everything. Your willingness to engage with the challenges and opportunities of quantum computing helps shape a more informed and inspired future.

This book would be incomplete without your interpretation, reflection, and continued exploration. May it serve as both a foundation and a springboard for your next steps—whether in academia, industry, education, or innovation.

We're all still learning. And in quantum science, every question leads to deeper understanding—and perhaps even to new realities.

Here's to the quantum future. Thank you for being part of it.

About the Author

Laura Lee is an author, web designer, and entrepreneur who weaves creativity and innovation into everything she does. Born and raised in a small town, Laura developed a love for storytelling and design at an early age. Her books often reflect her journey and the lessons learned while building her own digital design agency from the ground up. Laura's expertise in web design and online branding has made her a sought-after consultant for businesses eager to create meaningful connections in the digital world. Through her work, Laura inspires others to blend passion with purpose and design a life they love.

www.ingramcontent.com/pod-product-compliance
Lightning Source LLC
LaVergne TN
LVHW052225200926
848481LV00001B/215